Nanomaterial - Based Biomedical Applications in Molecular Imaging, Diagnostics and Therapy

Amitabha Acharya
Editor

Nanomaterial - Based Biomedical Applications in Molecular Imaging, Diagnostics and Therapy

Springer

Editor
Amitabha Acharya
Biotechnology Division
Institute of Himalayan Bioresource
Technology
Palampur, Himachal Pradesh, India

ISBN 978-981-15-4279-4 ISBN 978-981-15-4280-0 (eBook)
https://doi.org/10.1007/978-981-15-4280-0

This Springer imprint is published by the registered company Springer Nature Singapore Pte Ltd.
The registered company address is: 152 Beach Road, #21-01/04 Gateway East, Singapore 189721, Singapore

To my parents for raising me the way they did

Preface

With the advent of the nanotechnology, there is an enormous increase in the use of the fabricated nanoscale structures or nanomaterials (NMs) and devices. In recent years, the fabrication of NMs and its explorative properties has gained the attention of many scientific branches, namely, physics, chemistry, biology, engineering, computer science, etc. The primary interest in NMs arises from the fact that these are small enough to interact with cellular machinery and probably can reach previously inaccessible targets. Because of this dimensional property, NMs have been extensively used in biology or medicine, namely, for biological imaging, as drug and gene delivery system, and for both in vitro and in vivo applications, as fluorescent biological labels. NPs are promising therapeutic delivery vehicles which can host a wide range of active components, including chemotherapeutics, contrasting agents, proteins, nucleic acids, etc., for various biomedical applications. Nanoscale materials are progressively making a key impact on human health, are used increasingly more in diagnostic and therapeutic applications, and have the potential to revolutionize the future nanomedicine.

This book is focused toward the practical applications of medical nanotechnologies with a significant emphasis on specific examples from the existing recent literature. It offers a comprehensive review on application of the NMs for use in molecular imaging, disease diagnostics, therapy, and nanomedicine. It is written in an introductory as well as application level to allow the audiences to have a comprehensive knowledge and firm hold on the principles of NMs, nanotechnology, and nanomedicine, which is followed by a detailed discussion on the mainstream medical applications.

The contents of this book are systematically organized to emphasize the basic principle of nanotechnology, NMs, and their use in the field of biomedical applications like molecular imaging, diagnostics, and therapy. The facts that are unique to this book are (1) detailed description of the basic principles of the NM used in the field of molecular imaging, diagnostics, and therapy; (2) comprehensive discussion on the novel biocompatible NMs, their synthesis, and modern state-of-the-art characterization techniques; and (3) a foretaste into the future of nanomedicine and what we can evolve regarding NMs-based unique diagnostics and therapeutics. This book provides comprehensive outlook of the subject to wider audience, namely, biomaterials scientists/researchers, doctors, imaging researchers, biomedical scientists, postgraduates, and academicians.

Chapter 1 gives a brief introduction to the very new concept of personalized medicine. It covers the journey of the NMs and their advantages in personalized nanomedicine and briefs the potential future prospects of the biocompatible NMs in personalized medicine.

Chapter 2 discusses an overview in the area of molecular imaging, diagnostics, and therapy. It gives an emphasis on the importance and basic principles of different molecular imaging and diagnostic techniques.

Chapter 3 includes a detailed discussion on different methodologies applied for the advanced synthesis of biocompatible fluorescent NMs and their comprehensive characterization using different state-of-the-art techniques, followed by their detailed investigation as molecular imaging probes. It also gives a brief discussion on the biocompatibility of these NMs.

Chapter 4 provides the window for the advanced application of these NMs for personalized as well as on-site biomedical diagnosis. It covers the fundamentals of medical biosensors for point-of-care applications followed by materials and fabrication of medical biosensors to different technologies and operational techniques. Finally, it gives a brief overview on the current applications of this technology.

Chapter 5 discusses on different types of strategies that are used for the immobilization of the biomacromolecules on the NM surface, followed by their characterization using different spectroscopic and microscopic techniques. It also gives a brief detail on different methodologies employed for increasing the stability of these biomacromolecules on the NM surface and describes biomedical applications of these immobilized biomacromolecules on NMs in detail.

Chapter 6 gives a detailed information about the lectin nanoconjugates, their synthesis using different types of biocompatible (plant/non plant based) raw materials, and their systematic characterization, followed by different types of functionalization and conjugation strategies. It concludes with the comprehensive details on targeted therapeutic potential of these functionalized lectin nanoconjugates.

Chapter 7 presents different plant sources and methods applied for the synthesis of these biocompatible NMs, followed by their complete characterization by different spectroscopic and microscopic techniques. It focuses on recent advancements in this field with latest literature on their use in different biomedical applications.

Chapter 8 describes recent knowledge and research done in the field of biomolecule-NM interaction with the concept of biomolecular corona. Comprehensive discussion has been done on their formation, characterization, and application in the fields of molecular imaging, diagnostics, and therapy, etc. This chapter concludes with a section listing and describing naturally occurring complex, immune-mediated reactions (infusion reactions) occurring within the human body after receiving the NM (as a therapeutic dose or any other means).

The book concludes with a glimpse into the conceptual future of nanomedicine in Chapter 9. It includes the critical overview of the subject, current scenario, and future prospects of the NMs in biomedical applications. It further describes in enough detail future advancements in the field of NMs used in biomedical applications from leading research groups and scientists to capture and increase the student's interest and imagination about the potential and versatility of NMs for use in the field of molecular imaging, diagnostics, and therapy.

Palampur, Himachal Pradesh, India Amitabha Acharya

Contents

About the Editor

Amitabha Acharya is a Senior Scientist at the Nanobiology Laboratory, Biotechnology Division, CSIR-IHBT, Palampur. He graduated in Chemistry (Hons.) from the University of Calcutta, India, completed his postgraduate studies in Inorganic Chemistry at Banaras Hindu University, India, and received his Ph.D. degree in the area of Chemical Nanotechnology and Nanobiology from IIT Bombay under the supervision of Prof. C. P. Rao. His current research focuses on the synthesis and characterization of functional nanomaterials for molecular imaging, probe development, biomedical applications of nanomaterials against biofilm activity, the development of nano-chaperons as protein aggregation inhibitors, and understanding nanoparticle-cell/protein interactions. His contributions in the area include the development of novel carbon nanomaterials from Himalayan bioresources and the identification of key proteins responsible for protein corona formation under physiological conditions. He has published over 18 research articles and 4 review articles in peer-reviewed international journals and authored or co-authored 4 book chapters. He also holds three patents. Early in his career, he passed the GATE examination in India (AIR 129), and was a recipient of JRF and SRF fellowships from CSIR, India. He later received a Dr. D.S. Kothari postdoctoral fellowship. He has served as reviewer for several journals and is a life member of several scientific societies and organizations.

1

Nanomaterials: From Research to Personalized Medicine

Aqib Iqbal Dar and Amitabha Acharya

Contents

Abstract

The integration of nanotechnology with the personalized medicine has provided unparallel and unique opportunities for improving the way of treatment for many dreadful diseases. Nanomaterials (NMs), because of their explorative properties, offer a number of advantages in the therapeutics and diagnostics of diseases. The NM-based tools offer wide tunability in designing of the nanomedicine because of the smaller sizes, the large surface-to-volume ratio, and also the ease with which the NM surface can be modified to get desirable cell/tissue targeting. In this introductory chapter, a brief overview on the use of NMs for the personalized nanomedicine and the factors that need to be considered for successful translation of these personalized nanomedicines has been addressed. Also, a brief summary is given on different types of NMs that have been used for personalized medicine and are under clinical trials. Finally we have concluded the chapter with the

A. I. Dar · A. Acharya (✉)
Biotechnology Division, CSIR-Institute of Himalayan Bioresource Technology, Palampur, Himachal Pradesh, India

Academy of Scientific & Innovative Research (AcSIR), CSIR-Institute of Himalayan Bioresource Technology, Palampur, Himachal Pradesh, India
e-mail: amitabha@ihbt.res.in

A. Acharya (ed.), *Nanomaterial - Based Biomedical Applications in Molecular Imaging, Diagnostics and Therapy*, https://doi.org/10.1007/978-981-15-4280-0_1

perspective about how the proper understanding of nanobiotechnology, personalized medicine, and genomics can provide a better system for diagnosis, treatment, and management of diseases at the individual level.

Keywords

Personalized medicine · Nanomaterials · Genomics · Targeted delivery · Molecular profile

Abbreviations

CVDs	Cardiovascular disease
ECM	Extracellular matrix
FDA	Food and Drug Administration
MABs	Monoclonal antibodies
NPs	Nanoparticles
POC	Point of care

1.1 Introduction

Personalized medicine in simple terms can be defined as the healthcare strategy where prescription of specified therapies and treatments are given to the individuals that are most suitable for them. In personalized medicine, both the genetic and environmental factors are taken into consideration for successful therapies and treatments. The interventions of nanotechnology in the field of medicine has transformed into a new field of nanomedicine, which has created huge opportunities for significantly improving the dimensions for the treatment of many severe diseases (Jain 2009). With the swift increase in the growth of the nanomedicine field, the most promising innovations and applications can be found for personalized medicine. It is known that each drug is having differential effects on different types of individuals. Whether in terms of efficacy and safety of these drugs, these show differential behavior, because of the complexity and heterogeneity of the individuals (patient as well as disease). With proper and comprehensive understanding of genomics and proteomics and also the emergence of a plethora of novel and advanced technologies and patient-based molecular profiling, the potential of nanomedicine has knocked the doors of the future for the personalized medicine. This chapter includes a brief overview on the use of NMs for the personalized nanomedicine and the factors that need to be considered for successful translation of these personalized nanomedicines. Also, a brief summary is given on different types of NMs that have been used for personalized medicine and are under clinical trials. Finally we have concluded the chapter with an outlook about how the proper understanding of nanobiotechnology and personalized medicine can provide a better system of disease diagnosis; treatment at the individual level will probably fill the gaps between the patient's health and economical healthcare.

1.2 The Era of Personalized Medicine

The unique and the most innovative concept of personalized medicine has yet not been globally defined in terms of both the research and clinical perspectives. It could be defined as the management of disease of a patient or in other words the drug response utilization of the knowledge based on the clinical and molecular as well as bioinformatic outputs for providing the best possible treatment or therapy for a particular individual. With integration of personalized medicine with the nanotechnology and molecular outputs at the genomic, proteomic, and metabolomic level, it is possible to fill the gaps of personalized medicine. Gaining further knowledge will also improve the efficacy and acceptability of personalized medicine for specific individuals for evaluating the prognosis, therapy, and treatment and also monitoring throughout the clinical care of the patient (Vizirianakis and Fatouros 2012).

1.3 Personalized Nanomedicine: The Fusion of Genomics, Proteomics, and Metabolomics with Nanotechnology

The personalized medicine also called as precision medicine is the most advanced and modern way of integration of biotechnological and medicinal approaches. For decades, this approach has helped to decipher the mechanisms of pathophysiology of many diseases, their molecular diagnosis and therapeutics. Personalized medicine, as a new dimension of healthcare, combines the human-based big data analysis, their molecular profiling (genomics, metabolomics, and proteomics), and population health. This evidence-based healthcare is meant to increase the benefits and decrease harms that are generally observed for most of the medical treatments and therapies. So in this approach, the patient's individual profiles as well as their preferences are generally taken under consideration. In personalized medicine the medical treatments and therapies are adapted to the particular type of characteristics of each individual patient. In this approach, the most important and critical factors that are taken under consideration are the patient's genetic makeup; the environmental factors; their beliefs, attitudes, and knowledge; and also the social context. The precision medicine model of healthcare shows a great reliability on the data obtained from the patient, their analysis, and information. For adopting and implementing of precision medicine in the healthcare, this model goes way beyond the concept of genomics and has vast implication on the global health program. For elucidating the actual potential of the personalized medicine, it should hold the ecosystem of patient's attention and engagement, health, genomic and molecular profiling techniques, and also the analysis of big data. The emergence of personalized medicine in the twenty-first century is often regarded as one of the important and critical leap in the healthcare society, which possesses great potential to specifically diagnose and treat most of the dreadful disease present in the humankind. The personalized medicine always aims to customize the therapeutic process by incorporation of not only the data from ex vivo genetic and proteomic profiling, but to a greater extent, it entails the insights of in vivo imaging of disease type, its

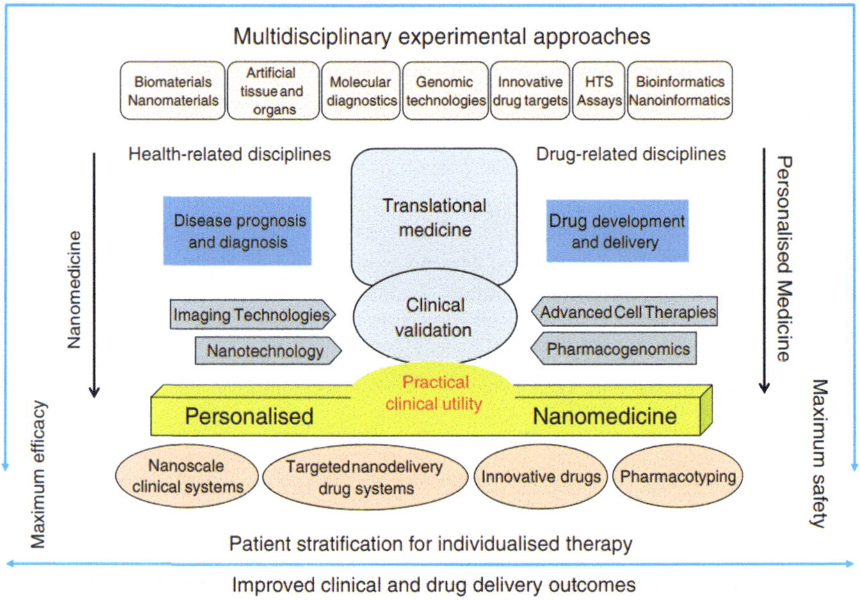

Fig. 1.1 An outline of the era of personalized medicine, describing different disciplines and experimental approaches. (Adapted from Vizirianakis and Fatouros (2012), reprint with permission)

stage, as well as the grade. The personalized medicine also relies on the responses of a particular patient to a particular type of treatment that is individually and specifically given at a particular time. The approach of personalized medicine relies in understanding of the person's unique genetic and molecular profile and how this makes a particular individual susceptible for a certain type of disease. It is also required to rethink and to change the idea of "one drug to fit all", which completely is in contrast with the present idea of personalized medicine. It helps in the management of patient's health considering the individual characteristics, viz., gender, age, height, weight, diet, and all other environmental factors. The approach of personalized medicine is a multidimensional approach that not only helps in improving the theranostic ability but also shows potential in early-stage disease detection, when the disease is easier to be treated effectively (Fig. 1.1).

The main interest in the concept of personalized nanomedicine comes from the fact that, while the personalized nanomedicine can lead us to the designing and applicability of the NMs, nanotechnology can also assist in the collection of data from molecular and genetic profiling of the individuals that are critical for the proper implication of personalized medicine. The recent advancements in the concept of personalized medicine have come into reality through proper development of NMs and also the technologies that could possibly help in early disease detection, their imaging, and also for specific identification of disease signatures at the molecular level. The excellent combinatorial approaches from the fields of pharmacology, genomics, proteomics, and metabolomics indeed have enabled us to successfully

Table 1.1 Examples of nanomedicines that are currently being found in the industry for the treatment of different diseases

Drug name	Active principle	Type of nanoformulation	Indication/disease	Status of clinical trial
Abraxane	Paclitaxel	Albumin NPs	Various cancers	Marketed
Patisiran	siRNA	Liposome	Transthyretin amyloidosis	Phase II
Ambisome	Amphotericin B	Liposome	Fungal infections	Marketed
Caelyx	Doxorubicin	PEGylated liposome	Solid tumors	Marketed
DaunoXome	Daunorubicin	Liposome	Solid tumors	Marketed
Doxil	Doxorubicin	PEGylated liposome	Various cancers	Marketed
Invega Sustenna	Paliperidone palmitate	Nanocrystal	Schizophrenia	Marketed
L-490	Insulin	Polymeric NPs	Diabetes type I	Phase I
Tricor	Fenofibrate	Nanocrystal	Hypercholesterolemia	Marketed
Rapamune	Rapamycin	Nanocrystal	Immunosuppressive	Marketed
Megace ES	Megestrol	Nanocrystal	Anti-anorexic	Marketed
Lipotecan	Camptotethin	Polymeric micelles	Various cancers	Phase I/II
NC-6004	Cisplatin	Micelle	Lung cancer	Phase I/II

and comprehensively investigate the complex molecular and genetic profiles of the patients. This scientific knowledge has offered mechanistic insights of various diseases and to correctly help in mitigating the therapeutics for the specific diseases. Nanotechnology has also been able to widely manipulate and expand the genetic and molecular profile-based information to be gathered from the patients. For instance, the use of GeneChip®-based microarrays deciphers the pattern of biomacromolecules present on the surface with proper control on their spatial confirmation to get the sequence of DNA at the nanoscale level (Ferrari 2005). The control over the deposition of molecules at the nanoscale level can aid in designing and increasing the density of information to several million folds for determination of the proteomic and genomic profiles (Lee et al. 2002, 2003). Another example is an FDA-approved gold NP modified with certain biorecognition molecules that were used for highly efficient genomic detection (Thaxton et al. 2009).

The application and integration of the nanotechnology to personalized medicine have provided an unparallel and excellent opportunity for the scientific and clinical world for improving the way of treatment of many diseases. The ability to manipulate and exploit the NMs has made it possible to develop more effective nanomedicines for the treatment of a number of diseases like cancers, CVDs, neurodegenerative diseases, etc. (Table 1.1). However, there are a number of challenges and obstacles that impede the process of personalization of a particular medicine. In spite of these challenges and hurdles, advanced efforts are made to overcome these challenges to develop more promising nanosystems that possibly will prove to be a potential and universal personalized medicine (Fornaguera and García-Celma 2017).

1.4 Challenges in Personalized Medicine

The personalized nanomedicine involving disease treatment and prevention consists of therapies specified to a particular individual or a group of individuals to achieve best possible outcomes. The birth of the concept of "omics," viz., pharmacoproteomics, pharmacogenomics, and pharmacometabonomics, marked the origin of personalized medicine in the early twentieth century. This enabled gaining inclusive knowledge of genetics of the individuals. With the growing advancements, a plethora of nanomedicines are present in the market, but none of them is actually designed as personalized medicine. For a personalized nanomedicine to become the medicine of the future, it should have certain advantages and salient feature (Table 1.2).

This dearth of current nanomedicine to be used as the personalized medicine can be accredited to a number of factors. One of the prime limitations is the issue that is always seen in the nanomedicine formulation. The present nanomedicines for therapeutic purpose are considered first-generation therapies without active tissue-targeting moieties that alters these medicines for targeting the specific type of cell or organ. This lack of active targeting ability can be considered as a formulation issue that needs to be corrected or avoided. For decades, doxil has been used as a potential anticancer drug for a number of tumors, but we must consider the fact that each region of tumor has a different microenvironment, viz., hypoxic gradient, fenestration, ECM, stroma cells, etc. Therefore, there is a critical need of excellent second-generation nanomedicines and therapies that could actively and specifically target the tissue of interest precisely. A number of actively targeting moieties have been used for this purpose, but the most important and efficiently used are monoclonal antibodies (MABs) that target single cells containing a specific expressed moiety

Table 1.2 Advantages and features of nanomedicines to become the medicine of the future and the properties for application in personalized nanomedicine

S.no.	Properties of personalized nanomedicine
1.	The medicine should be in the nanometric-scale range.
2.	The medicine should have wide tunability and versatility.
3.	There should be possibility of using the labile compounds like the siRNA.
4.	The active principles should be encapsulation and protection.
5.	The nanomedicine should have presence of specific organ-targeting moieties.
6.	It should be able to enter to specific cellular compartments.
7.	It should be able to adapt with the requirements of a patient group.
8.	It should have a definite adaptation of dosage, frequency, etc. for patients.

(Béduneau et al. 2008). The other classes of widely tunable nanosystems that have potential in personalized nanomedicine market are the bacteriophages and viruses (Sunderland et al. 2017). These types of nanomedicines are modified genetically for increasing the expression or suppression of gene of interest to tackle a particular disease. For nanomedicine personalization, the use of genetic material as an active principle of the nanomedicines is generally recommended compared to most of the pharmacology-based therapies using conventional drugs. So, there is a need of critical and thorough research, directed toward gene therapies rather than using present nanomedicines.

Commercial Issues Before the successful entry of the medicine, the commercial and economic issues are the most important and challenging that need to be considered. The entire commercialization process for a novel drug to reach to the market is estimated to take almost 10–15 years, with a cost of ~ $1 billion, and more in case of personalized medicines (Bobo et al. 2016). So, the overall process of personalization seems nonrealistic and also less efficient for providing a true therapeutic solution to the patients at the individual level considering the present scenario of technologies. Hence, it will be a prime and critical research objective for the scientific world to identify groups of individuals that will fit into the common parameters and characteristics to develop personalized therapies for them.

As the outcome, the comprehensive and extensive research on the personalized nanomedicine must address the molecular profiles of individuals, describing the characteristics of genetic expression. More importantly, it should also address the pattern of characteristics that are expressed or suppressed over a period time and also throughout the progression of a particular disease. By acquisition of the basic knowledge, the clinicians could develop gene-specific nanotherapies by defining the stages and progression of a particular disease at a definite time course (Fig. 1.2).

1.5 Limitations and Hurdles in Personalized Nanomedicine

Though personalized nanomedicine is a promising approach for the disease treatments and therapies, there are a number of challenges that are associated with this approach. For making the approach of personalized medicine more potential and universal, these challenges need to be overcome. Due to the defined and novel nanomolecular interactions, the manipulation of the NMs is rather difficult and complex to be executed. These effects complicate the utility, reliability, and also the predictability of the nanomedicine. Further, there are potential risks in designing and formulating the nanomedicine that makes it difficult for clearing the clinical ethics for assessing its clinical utility. The rules and regulations for approving a particular type of nanomedicine product can be considered as one of the critical obstacle in the way of personalization of the medicine. This can be related to the fact that most of the personalized treatment approaches are inherently and safely designed for a particular individual, but not for the population. The integration of

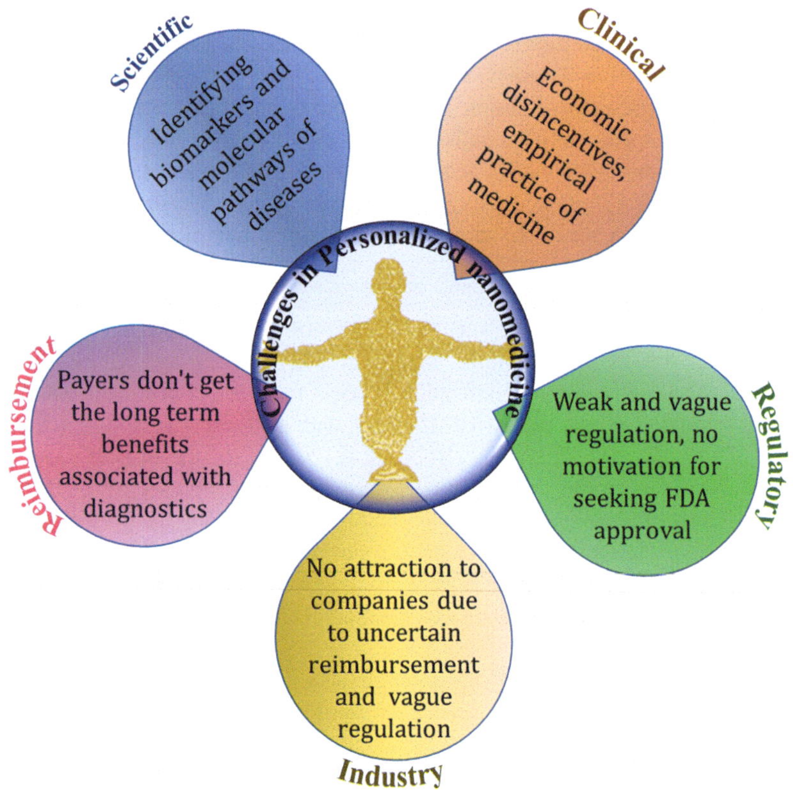

Fig. 1.2 Challenges and limitation of personalized medicine at the different levels that affect proper translation and commercialization

NMs into the approach of personalized medicine creates a buzz because of their ethical issues. The collection of huge data is done from the nanodiagnostics, and genetic evaluation of individuals, particularly point of care (POC) nanodevices that may bypass the usual healthcare strategies, will greatly affect the collection of data on the individual basis. These big molecular data collections possibly will challenge the healthcare society in respect of storing and handling of the data as well as the privacy-related issues. Another major obstacle in the way of personalized medicines is the heterogeneity in the genetic testing within the similar individuals as well as the diseased tissues or organs (Gerlinger et al. 2012). All these limitations and obstacles are challenging for the universality and applicability of the personalized medicine.

1.6 Conclusions and Perspectives

Although the concept of personalized nanomedicine is yet not fully understood and is in the initial stages of development, this approach is emerging and gaining more and more acceptance in the present world. It has become an independent research field and more integrated with the nanotechnology for further improving the acceptance at the individual level of treatment and therapies. With the advent of personalized nanomedicine, it has given a significant stimulus to the disciplines of medicine and pharmacy, for providing advanced clinical treatments, disease management, diagnosis, and drug delivery. Thus we believe that it is the prime time to organize and cultivate the model of personalized medicine for exploiting the knowledge achieved from nanotechnology, genomics, proteomics, and metabolomics. Also, there are a number of nanomedicines that are already going in different phases of clinical trials. A critical evaluation of the clinical needs integrated with the better knowledge of NM's biophysicochemical parameters will accelerate the development of safe and effective personalized nanomedicine. Therefore, the vision for the future should be the advancement in the translation of personalized nanomedicine to every possibly disease.

Time by time, we will get more and more closer to the advancements of personalized medicine by developing more crucial and intricate interactions between the nanotechnology and "omics." Thus, there will be establishment of more reliable personalized nanomedicine for healthcare of patients. Further, by revealing molecular and genetic heterogeneity, we will better understand the pharmacological response and its variability among the individuals. Thus, strengthening and providing better facilities of translational medicine to the patients and a greater utility of the efforts of personalized nanomedicine will be accomplished.

Acknowledgments The authors would like to thank the director of CSIR-IHBT for his constant support and encouragement. AA acknowledges the financial support from CSIR (MLP-201) and DST (GAP-0214; EMR/2016/003027). AID thanks DST for providing SRF. The CSIR-IHBT communication number of this manuscript is 4547.

References

Béduneau A, Hindré F, Clavreul A, Leroux JC, Saulnier P, Benoi JP (2008) Brain targeting using novel lipid nanovectors. J Control Release 126(1):44–49

Bobo D, Robinson KJ, Islam J, Thurecht KJ, Corrie SR (2016) Nanoparticle-based medicines: a review of FDA-approved materials and clinical trials to date. Pharm Res 33(10):2373–2387

Ferrari M (2005) Cancer nanotechnology: opportunities and challenges. Nat Rev Cancer 5 (3):161–171

Fornaguera C, García-Celma M (2017) Personalized nanomedicine: a revolution at the nanoscale. J Pers Med 7(4):12–31

Gerlinger M, Rowan AJ, Horswell S, Larkin J, Endesfelder D, Gronroos E, Martinez P, Matthews N, Stewart A, Tarpey P, Varela I, Phillimore B, Begum S, McDonald NQ, Butler A, Jones D, Raine K, Latimer C, Santos CR, Nohadani M, Eklund AC, Spencer-Dene B, Clark G, Pickering L, Stamp G, Gore M, Szallasi Z, Downward J, Futreal PA, Swanton

C (2012) Intratumor heterogeneity and branched evolution revealed by multiregion sequencing. N Engl J Med 366(55):883–892
Jain KK (2009) Role of nanobiotechnology in the development of personalized medicine. Nanomedicine 4(3):249–252
Lee KB, Park SJ, Mirkin CA, Smith JC, Mrksich M (2002) Protein nanoarrays generated by dip-pen nanolithography. Science 295(5560):1702–1705
Lee KB, Lim JH, Mirkin CA (2003) Protein nanostructures formed via direct-write dip-pen nanolithography. J Am Chem Soc 125(19):5588–5589
Sunderland KS, Yang M, Mao C (2017) Phage-enabled nanomedicine: from probes to therapeutics in precision medicine. Angew Chem Int Ed 56(8):1964–1992
Thaxton CS, Elghanian R, Thomas AD, Stoeva SI, Lee JS, Smith ND, Schaeffer AJ, Klocker H, Horninger W, Bartsch G, Mirkin CA (2009) Nanoparticle-based bio-barcode assay redefines "undetectable" PSA and biochemical recurrence after radical prostatectomy. Proc Natl Acad Sci U S A 106(44):18437–18442
Vizirianakis IS, Fatouros DG (2012) Personalized nanomedicine: paving the way to the practical clinical utility of genomics and nanotechnology advancements. Adv Drug Deliv Rev 13 (64):1359–1362

Introduction to Molecular Imaging, Diagnostics, and Therapy

2

Avnesh Kumari, Pooja, Sarthak Sharma, and Amitabha Acharya

Contents

Abstract

Nanotechnology has given a plethora of nanoparticles (NPs) during the last decades, which have been extensively used in molecular imaging, diagnostics and therapy. Despite recent progress in this area, there are only a few

A. Kumari · A. Acharya (✉)
Biotechnology Division, CSIR-Institute of Himalayan Bioresource Technology, Palampur, India

Academy of Scientific & Innovative Research (AcSIR), CSIR-Institute of Himalayan Bioresource Technology, Palampur, India
e-mail: amitabha@ihbt.res.in

Pooja · S. Sharma
Biotechnology Division, CSIR-Institute of Himalayan Bioresource Technology, Palampur, India

A. Acharya (ed.), *Nanomaterial - Based Biomedical Applications in Molecular Imaging, Diagnostics and Therapy*, https://doi.org/10.1007/978-981-15-4280-0_2

nanoformulations that have entered into clinical trials. We must identify new nanoimaging probes and their targets for therapy, develop new nanoagents for diagnostics, and position them in a clinically useful and cost-effective manner. This chapter summarizes the current status of NPs used as molecular imaging probes, their use in the diagnosis of disease-related analytes, and also their effective use in therapy.

Keywords

Bioimaging · Diagnostic · Drug delivery · Nanotechnology · Therapy

Abbreviations

ATP	Adenosine triphosphate
AuNC	Gold nanoclusters
GST	Glutathione S-transferase
HBV	Hepatitis B virus
MMP	Matrix metalloproteinases
NSE	Neuron-specific enolase
PSA	Prostate-specific antigen

2.1 Introduction

Recently there have been noteworthy developments in the use of imaging techniques to identify and monitor diseased tissues. Pathological and physiological changes can be identified through the visualization of tissue morphology and cell function (Minchin and Martin 2010). The rapidly developing field of molecular imaging boosted premature disease imaging and disease detection and helps in image-based healing and personalization of treatment. In addition, it also provides vital information on the therapeutic efficacy. Nonetheless, molecular imaging employs imaging modalities to capture and differentiate biological processes at the molecular and cellular level (Bao et al. 2013; Cormode et al. 2009; Chen et al. 2014; Weissleder and Pittet 2008). Recent advancements in the field of nanotechnology have promoted the growth of a variety of nanoparticles for molecular imaging and diagnostic purposes. Nanoparticulate probes have confirmed significant advantages over single molecule–based contrast agents. NPs furnish magnetic, radioactive, and optical modalities to bring about increased contrast in a variety of molecular imaging protocols. An extra benefit of NPs is that manifold modalities can be tagged in identical NPs in respect to permitting them as a functional multimodal imaging probe, which can improve vital facts for disease diagnosis (Chen et al. 2016a, b). Diagnostic NPs intend to capture pathologies and advance the grasping of key pathophysiological basis of a range of ailments and ailment treatments. Nanodiagnostics are only helpful in a limited instances, due to their intricate pharmacokinetic properties (Baetke et al. 2015). NP-based sensing is the main force for medical diagnostics. NPs offer a huge surface-to-volume ratio to allow binding of a great number of biomarkers to detain the analyte, thus allowing extremely responsive detection (Chen et al. 2016a, b). In vitro diagnostics uses blood, urine, sweat, sputum, and other body fluids collected from diseased/normal humans for detection of biomarkers, which are signals for improper functioning of the body, and can take part in diagnosis of human diseases (Prasad 2004). Many techniques such as fluorescence, absorption, and Raman can be used to compute the properties and fate of NPs in various applications (Chen et al. 2016a, b). Therefore, the bulk of NPs formulations at present aim to increase the gathering and release of therapeutic entity at the pathological spot, augment curative efficiency, and decrease the occurrence and strength of after effects in normal tissues (Kiessling et al. 2014; Duncan and Gaspar 2011; Rizzo et al. 2013). The inherent characteristics of NPs hold immense potential for imaging, diagnostic, and therapeutic agents into a solitary NP formulation. Such formulations find use in imaging diseased tissues, monitoring the biodistribution and target site accumulation, measuring drug discharge, and longitudinally evaluating the therapeutic activity. Such NPs may be used for developing personal nanomedicine-based treatments by allowing patient preselection and by managing therapeutic activity (Lammers et al. 2011; Lammers et al. 2012; Theek et al. 2014). This chapter covers the latest developments in NP formulations for molecular imaging, diagnostics, and therapy.

2.2 Molecular Imaging

Molecular imaging is a noninvasive technique which focuses at sensing exact molecular targets and basic biological processes (Jaffer et al. 2007). Molecular imaging requires accurate and highly specific imaging agents. Such imaging agents have two important features, namely a signal detection compound and an affinity ligand, that recognizes anticipated molecular targets (Jaffer et al. 2007). Diseases have well-established targets of biological and clinical importance, as well as targets which have signal amplification potential such as enzymes and receptors (Jaffer et al. 2007). Usually, there are two categories of imaging methods—morphological and functional or molecular imaging techniques (Padmanabhan et al. 2016). Morphological imaging techniques include ultrasound (US), computed tomography (CT), and magnetic resonance imaging (MRI) that are performed by clinical translation and high spatial resolution. Molecular imaging techniques such as optical imaging (OI), single-photon emission computed tomography (SPECT), and positron emission tomography (PET) are carried out by probes with high sensitivity (Padmanabhan et al. 2016). However, all of these methods have their own boundaries and require their individual customized probes (Padmanabhan et al. 2016). Nanomaterials (NMs) have played noteworthy function in molecular imaging. NMs have new electronic, structural, optical, and magnetic properties, which make them suitable for in vitro and in vivo applications. NMs have been extensively used in molecular imaging. NMs used for molecular imaging provide brightness, inertness to their microenvironment, and a more even distribution (Wolfbeis 2015). NMs do not undergo nonspecific binding by cellular biomacromolecules, and they are noncytotoxic. Fluorescent NMs do not undergo specific binding with proteins of biological milieu and also their optical properties are not affected by the proteins (Wolfbeis 2015). Fluorescent silica nanoparticles, quantum dots, nanoclusters, and upconversion nanoparticles have been extensively used in molecular imaging.

2.2.1 Fluorescent Silica Nanoparticles

Silica nanoparticles (NPs) were among the first NMs to be used in molecular imaging. Different types of silica NPs synthesized by various methods have been extensively used in bioimaging. Among these, PEG-coated silica NPs possess single pore, sizes of 9 nm, and uniform size distributions. Among these, mesoporous silica NPs (MSNs) are more commonly used because they can be loaded with fluorescent dyes, photosensitizers, or diagnostic agents (Wolfbeis 2015). Silica NPs can be easily conjugated with organic dyes to form imaging modality. Organic dyes can be conjugated to silica matrix (Ahn et al. 2013; Huang et al. 2011a, b; Liong et al. 2008; Lu et al. 2009) or pores on the surface (Ciccione et al. 2016; Heidegger et al. 2016; Pan et al. 2013). FITC-conjugated amine-modified MSNs showed concentration-dependent uptake in mouse splenocytes under fluorescent microscopy (Heidegger et al. 2016). In another study, in vivo biological behavior FITC-conjugated MSNs of various size and shapes were investigated by intravenous injection to mice. It was found that different shaped MSNs showed differences in biodistribution. For example, the distribution of rod-shaped MSNs in the spleen, while entrapment of small rod–shaped MSNs in the liver (Huang et al. 2011a, b). Encapsulation of indocyanine green (ICG) molecules in MSNs showed better fluorescence than non-encapsulated ICG molecules. Results suggested that these NPs have better photostability as compared to the dyes in the free form. Moreover, these MSNs provided much effectiveness in fluorescence-based cell/tissue tracking under in vivo environmental conditions (Lee et al. 2009).

2.2.2 Quantum Dots

Quantum dots are semiconductor fluorescent nanocrystals that have very high quantum yields and show better photostability and biocompatibility compared with organic dyes (Kobayashi et al. 2009; Michalet et al. 2005; Zhang et al. 2008). They consist of any semiconductor material with organic coating. They have also been used for molecular imaging of cancer as they permit segregation of signal generated by QD from automatic fluorescence of tissue because of huge stokes shift and extended fluorescence period (Tokumasu and Dvorak 2003; Ghazani et al. 2006; Nisman et al. 2004). Antibody-conjugated QDs have been successfully reported in the detection of QD-labeled biomarkers with good resolution (Giepmans et al. 2005). Another study reported use of bioconjugated quantum dot for multiplexed labeling and quantification of five clinically important markers for breast cancer, namely,

HER2, ER, PR, EGFR, and mTOR (Yezhelyev et al. 2007). Similar study reported that QD–streptavidin bioconjugates showed significantly higher signal to noise ratio in amplification of clinically important ERBB2 gene (Xiao and Barker 2004).

2.2.3 Nanoclusters

Nanoclusters are a new class of materials with only a few to tens of atoms and the size of their cores is usually below 2 nm (Shang et al. 2013). They have unique optical and electrical properties, for example, highest occupied molecular orbital to lowest unoccupied molecular orbital transition, tunable fluorescence, and large Stokes shift (Shang et al. 2013). D-penicillamine conjugated with gold nanoclusters showed small size, high colloidal stability, and brightness and have been employed for biological imaging of human cancer (HeLa) cells (Shang et al. 2011a, b). The dihydrolipoic acid (DHLA)–gold nanoclusters (AuNCs) have been used for exploring the fluorescence lifetime imaging in HeLa cells. HeLa cells exposed to DHLA–AuNCs exhibited a long fluorescence lifetime of 500–800 ns (Shang et al. 2011a, b). Folic acid–conjugated trypsin-stabilized gold nanoclusters have been used for in vivo cancer fluorescence imaging (Liu et al. 2013a, b). Ovalbumin-protected gold nanoclusters conjugated with folic acid (FA-Ova-AuNCs) were employed for specific staining of HeLa cells and for the detection of cancer through cell imaging (Qiao et al. 2013). Transferrin (Tf)-functionalized gold nanoclusters (Tf-AuNCs)/graphene oxide (GO) nanocomposite (Tf-AuNCs/GO) based turn-on near-infrared (NIR) fluorescent probe was used for imaging of cancer cells. The fluorescent probe was absorbed by HeLa cells, and remarkable fluorescence was seen after 4 h of incubation (Wang et al. 2013). Metallothionein-coated silver nanoclusters (AgNCs) possessed a good antioxidant capacity and showed intense fluorescence in HeLa cells (Hu et al. 2016). Folate-conjugated thiol polyethyleneimine-stabilized silver NCs (SH-PEI-AgNCs) with ultra-small size, low toxicity, intense NIR fluorescence, excellent photostability, and chemical stability hold great promise for in vitro and in vivo cancer imaging (Wang et al. 2014a, b).

2.2.4 Upconversion Nanoparticles

Upconversion nanoparticles (UCNPs) upconvert NIR light to multiwavelength light using a low-power continuous-wave diode laser. Excitation of NIR allows intense tissue infiltration and circumvents the automatic fluorescence of the living samples. They have exclusive properties such as high stability, low cytotoxicity, superior photostability, and photo-bleaching, which make UCNPs as exclusive ocular tools for biological studies (Liu 2011; DaCosta et al. 2014; Dong et al. 2015; Prodi et al. 2015; Bünzli 2016). $NaYF_4$:Yb^{3+}, Er^{3+} NPs covered with α-cyclodextrin have been evaluated as photoacoustic imaging agents. The in vivo confinement of the UCNPs in mice suggested these NPs as photoacoustic capturing agents (Maji et al. 2014). Antibody-conjugated $NaYF_4$:Yb/ErNPs have been applied for imaging HeLa cells with antigen expressed on the cell membrane and very precise staining of HeLa cells (Wang et al. 2009). In another study, cell imaging was demonstrated using Y_2O_3:Er NPs customized with cyclic arginine–glycine–aspartic acid peptides. They showed that these UCNPs can purposely bind to cancer cells with enhanced integrin $\alpha_v\beta_3$expression (Zako et al. 2009). Y_2O_3:Yb/Er UCNPs, ranging 50–150 nm, have been inoculated on nematode *Caenorhabditis elegans*, and the digestive system of the worm showed statistical distribution of UCNPs when excited at 980 nm (Lim et al. 2006). PEG polymer–coated Y_2O_3:Yb/Er UCNPs have been used for in vivo vascular imaging of nude mice. The PEG coating minimized nonspecific tissue binding and enhanced the half-lives of the UCNPs in the blood (Hilderbrand et al. 2009).

2.3 Diagnostics

The detection, identification, and quantitative investigation of biomarkers at very small concentrations for diagnosis of diseases represent a new border horizon in biomedical research. Gold nanoparticles, quantum dots, and carbon dots have been extensively used in diagnostic applications.

2.3.1 Gold Nanoparticles

Gold nanoparticles (AuNPs) have interesting application in molecular diagnostics field. They have been extensively used in diagnostics application due to their tunable optical properties and excellent biocompatibility (Radwan and Azzazy 2009). Colorimetric biosensor based on oval-shaped AuNPs have been used for detecting breast cancer cell lines (Lu et al. 2010). In another study, target sequence of 2.6 nM *Mycobacterium tuberculosis* extracted from patients was detected easily using unmodified AuNPs and single-stranded detection oligonucleotides (Tsai et al. 2013). Multiple antigenic peptide-coated AuNPs detected bluetongue virus (BTV) disease specific antibodies in the sera (Schwartz-Cornil et al. 2008). Heparin plays a key role in hemorrhage and thrombocytopenia throughout the anticoagulant therapy and surgery. Heparin detection was carried out using cysteamine-modified AuNP (cyst-AuNPs) and gold nanoclusters stabilized with trypsin (tryAuNCs) (Liu et al. 2013a, b). AuNPs coated with three short dye-terminated reporter sequences specific for binding with three mRNA transcripts: c-myc mRNA, TK1 mRNA, and GalNAc-T mRNA. These AuNPs when incubated with MCF-7 cancer cells and MCF-10A normal cells detected all the three mRNAs specific for breast cancer (Li et al. 2012a, b). Another study used AuNPs bound with three hairpin sequences labeled with fluorescein, Texas red, and Cy5. Such AuNPs hold enormous potential in multiplexed recognition of single base incompatible DNA targets (Song et al. 2009) (Table 2.1).

Table 2.1 Gold nanoparticle–based diagnostic assays with targets

Nanomaterial	Sample	Target	Limit of detection	Reference
AuNCs	Cell lysate	GST	25 μM	Chen et al. (2009)
AuNCs	Cell lysate	ATP	43 μM	Li et al. (2012a, b)
AuNCs	Serum	Urea	1 mM	Nair et al. (2013)
CaCO3/AuNC	Serum	NSE	2.0 pg/mL	Peng et al. (2012)
AuNPs	Serum	PSA	0.032 pg/mL	Liu et al. (2013a, b)
AuNPs	Tumor tissue	MMP	1 nM	Lee et al. (2008)
AuNP-QD	HepG2	HBV	1 virus particle/μL	Draz et al. (2012)
AuNP-QD	Malignant tissues	MMP-7	126 ng/mL	Kim et al. (2008)
Apo-GOx@AuNP	Cell lysate	Glucose	5 nM	Li et al. (2013)
Try-AuNCs	Serum	Heparin	0.05 μg/mL	Liu et al. (2013a, b)
Ova-AuNCs	Rat brain microdialysate	Glucose	0.10 mM	Wang et al. (2014a, b)
AuNCs	Human plasma	Pyrophosphate	28 μM	Li et al. (2012a, b)

2.3.2 Quantum Dots

miRNA dysregulation plays a very key function in the growth and development of cardiovascular and cancer diseases. Förster resonance energy transfer (FRET) between a luminescent Tb complex and semiconductor quantum dots (QDs) have been used to detect miRNA biomarkers in very low concentrations with better sensitivity (Qiu and Hildebrandt 2015). FRET from luminescent terbium complexes (LTCs) to semiconductor quantum dots was used for detecting five biomarkers with a better sensitivity, that is, 40–240-fold higher than well-established single-analyte reference assays (Geißler et al. 2010). QD-based FRET bioassay was used to detect thrombin activity in low volumes of serum and blood samples on paper strips covered within PDMS/glass sample chips (Petryayeva and Algar 2015). Another QD–FRET-based assay used paper-based solid-phase QD–FRET nucleic acid hybridization with green-emitting QD donors and Cy3 dye acceptors and showed the limit of detection (LOD) up to 450 fmol (Noor and Krull 2014). Multicolor QD–antibody conjugates have also been used to simultaneously detect a panel of four protein biomarkers characteristic of Hodgkin's lymphoma on human tissue biopsies. Results showed distinct QD staining pattern (CD15 positive, CD30 positive, CD45 negative, and Pax5 positive), which can be used not only to spot Hodgkin's lymphoma but also to distinguish it from benign lymphoid hyperplasia (Liu et al. 2010). Detection of genetic biomarkers of different diseases such as syphilis, HIV, malaria, hepatitis B and C was carried out using QD barcodes by detecting the genetic fragments of these pathogens. This can be performed within less than 10 min and requires a very low volume of sample having detection limit in the femtomol range (Giri et al. 2011).

2.3.3 Carbon Dots

Carbon dots (CDs) are ideal candidates for diagnostics due to their simplistic surface functionalization, biocompatibility, low cytotoxicity, excellent stability against photo bleaching, and their high aqueous solubility (Pirsaheb et al. 2019). CD, gold nanorod, and silica nanohybrids (GNR-SiO2-CD) have been used for successful detection of macrophages, which mimic unstable atherosclerotic plaques (Liu et al. 2018). Another study used tomato juice synthesized CDs for finding carcinoembryonic antigen (Miao et al. 2016). FRET-sensing platform using CDs and MnO_2 nanosheets was used for sensitive detection of microRNA. This platform was utilized to spot MCF-7 breast cancer cells and the spiked serum sample (Mohammadi and Salimi 2018). Another study used FRET nanoprobe consists of two CDs derived from resorcinol and citric acid along with MoS_2 nanosheets for the recognition of OVCAR-3 and MCF-7 cells lines (Hamd-Ghadareh et al. 2018). CA15-3 antibody coated CDs and AuNPs tagged PAMAM-dendrimer based FRET assay demonstrated good precision in the analysis of serum samples and MDA-MB-231 cancer cells (Mohammadi et al. 2018). Synthesis of CDs from D-glucose and L-aspartic acid localized specifically into glioma sites with much superior strength than normal brain tissue. Results documented that CD–Asp can be utilized as a targeted fluorescence imaging agent of brain glioma (Zheng et al. 2015). Recent study used AuNPs–CDs conjugate for the detection of α-L-fucosidase characteristic of hepatocellular carcinoma (Mintz et al. 2018).

2.4 Therapy

The main application of NPs in medicine is for therapy of diseases (Curry et al. 2014). NPs enhance the solubility, bioavailability, and therapeutic efficacy of drugs. Nanopharmacotherapy deals with safe use of drugs in patients, while reducing side effects and maximizing health benefits (Prasad 2012). Nanopharmacotherapy is affected mainly by two factors, namely pharmacokinetics (PK) and pharmacodynamics (PD). Nanopharmacotherapy has led to noteworthy enhancements in the PK/PD profile of drugs (Müller et al. 2001; Paxton 1981). Various therapies offered by NPs are drug delivery, photodynamic therapy, photothermal therapy, and antimicrobial therapy.

2.4.1 Drug Delivery

NPs have been extensively used for improving the performance of drugs. Different type of NPs made from biodegradable polymers such as PLGA, PLA, BSA, chitosan, gelatin, etc. have used for improving therapeutic efficacy of drugs (Kumari et al. 2010). Many nanoformulations are already in the market after passing clinical trials (Kim et al. 2004; Moen et al. 2009; Kamaly et al. 2012; Raj et al. 2013). Recently many inorganic NPs have shown potential in drug delivery applications. They include mesoporous silica NPs (Li et al. 2003; Zhang et al. 2014), quantum dots (Probst et al. 2013), AuNPs (Agasti et al. 2009), iron oxide NPs (Cheng et al. 2014a, b), etc. However, despite enormous publications in the above area, no formulation of inorganic NPs has entered into clinical trials (Chen et al. 2016a, b).

2.4.2 Photodynamic Therapy

Photodynamic therapy (PDT) is based on the creation of light-absorbing molecules identified as photosensitizers (PSs). Upon light irradiation at a precise wavelength, PSs create reactive oxygen species (ROS), which are lethal to the diseased cells. Presently, PDT is being broadly used in clinics for cancer therapy of the skin, pancreas, breast, prostate, and lung (Dolmans et al. 2003). Organically modified silica (ORMOSIL) and mesoporous silica NPs have been widely used for PDT due to their tunable size, shape, porosity, and dispersibility. Water-soluble HPPH was successfully encapsulated in ORMOSIL NPs. Upon irradiation, encapsulated PS was found to generate $^1O_2$ (Roy et al. 2003). In another study by the same group, another PS iodobenzylpyropheophorbide was loaded in ORMOSIL NPs. Again the encapsulated PS could retain its properties and released $^1O_2$ upon irradiation (Ohulchanskyy et al. 2007). Quantum dots (QDs) are promising nanomaterials which can be used in PDT due to their size and unique optical and emission properties. Hybrid nanocomposite of QDs with porphyrin (PS) and meso-tetra (4-sulfunatophenyl) porphyrin was found to generate $^1O_2$ upon excitation at 355 nm (Shi et al. 2006). In another study, FRET efficiency and generation of $^1O_2$ was improved by conjugating Rose Bengal (RB) and Ce6 peptide to QDs (Tsay et al. 2007).

2.4.3 Photothermal Therapy

Photothermal therapy (PTT) uses photothermal agents to achieve selective heating of the local environment. PTT agents absorb light, electrons make transitions from the ground state to the excited state, and excited state relaxes through emission of nonradiative decay. This results in increase in kinetic energy and overheating of local environment around PTT agents. This heat is employed for the destruction of tumor tissues (He and Bischof 2003). Currently, gold nanoshells, gold nanoparticles, gold nanorods, and gold nanocages are the main nanostructures which are used in PTT (Hainfeld et al. 2004;; Zharov et al. 2005; El-Sayed et al. 2006; Huang et al. 2006; Takahashi et al. 2006a, b; Huff et al. 2007; Hirsch et al. 2003; Loo et al. 2004, 2005; Chen et al. 2005a, b). This topic is covered in detail in review by Huang et al. (2008).

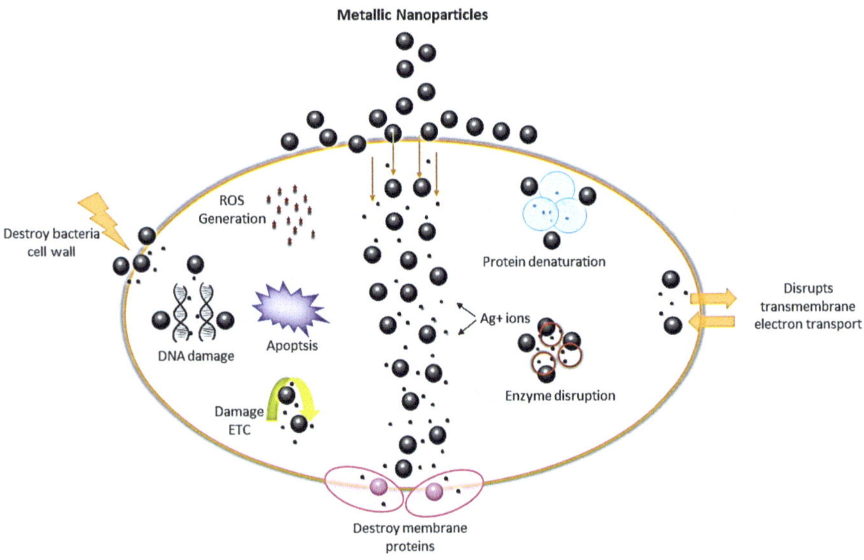

Fig. 2.1 Schematic representation of antimicrobial action of metallic nanoparticles

2.4.4 Antimicrobial Therapy

Metallic NPs, such as gold, silver, and copper, have shown activities against broad range of bacteria (Huang et al. 2011a, b; Huh and Kwon 2011; Zazo et al. 2017). The degree of antimicrobial activity of metallic NPs rely on physiochemical properties, namely size, shape, or surface charge (Pelgrift and Friedman 2013; Cheng et al. 2014a, b; Bai et al. 2015). It has also been observed that conjugation of antibiotics with NPs increases the antibacterial efficacy of antibiotics (Ahangari et al. 2013) (Fig. 2.1)

2.5 Conclusions

NPs due to their size, shape, surface area, and optical and magnetic properties are ideal candidates for molecular imaging, diagnostics, and therapy. Different types of NPs used in molecular imaging, diagnostics, and therapy are discussed in this chapter. NPs can be smartly designed for integrating imaging probes and drugs into single entity. Such NPs can be used for early disease diagnosis, detection of biomarkers with enhanced sensitivity, and curing of disease by release of drugs at the drug site. New multifunctional NPs and superior treatment methods are necessary to fulfill the requirements of real-time, noninvasive imaging for suitable drug delivery and therapy efficiency in vivo. Such NPs should have the capability to execute multiplex imaging that will allow us to develop a complete perceptive of tissue properties, molecular properties, and targeting mechanisms. Effort is also required to lessen the lethal effects of NPs and design safe NPs. Pains and hurdles will help to invent simple NPs that give secure and revealing images which meet the requirements of the regulatory bodies. This will also be supportive in the premature diagnosis and helpful in management of human diseases.

Acknowledgments The authors would like to thank Director of CSIR-IHBT for his constant support and encouragement. AA acknowledges the financial support from CSIR (MLP-201) and DST (GAP-0214; EMR/2016/003027). AK would like to thank CSIR-IHBT for support. Pooja and SS would like to acknowledge CSIR for their financial assistance fom MLP-0145 and MLP-0141, respectively. The CSIR-IHBT communication number of this manuscript is 4564.

References

Agasti SS, Chompoosor A, You CC, Ghosh P, Kim CK, Rotello VM (2009) Photoregulated release of caged anticancer drugs from gold nanoparticles. J Am Chem Soc 131(16):5728–5729

Ahangari A, Salouti M, Heidari Z, Kazemizadeh AR, Safari AA (2013) Development of gentamicin-gold nanospheres for antimicrobial drug delivery to Staphylococcal infected foci. Drug Deliv 20(1):34–39

Ahn B, Park J, Singha K, Park H, Kim WJ (2013) Mesoporous silica nanoparticle-based cisplatin prodrug delivery and anticancer effect under reductive cellular environment. J Mater Chem B 1 (22):2829–2836

Baetke SC, Lammers TGGM, Kiessling F (2015) Applications of nanoparticles for diagnosis and therapy of cancer. Br J Radiol 88(1054):20150207

Bai X, Li L, Liu H, Tan L, Liu T, Meng X (2015) Solvothermal synthesis of ZnO nanoparticles and anti-infection application in vivo. ACS Appl Mater Interfaces 7(2):1308–1317

Bao G, Mitragotri S, Tong S (2013) Multifunctional nanoparticles for drug delivery and molecular imaging. Annu Rev Biomed Eng 15:253–282

Bünzli JCG (2016) Lanthanide light for biology and medical diagnosis. J Lumin 170:866–878

Chen J, Wiley B, Li ZY, Campbell D, Saeki F, Cang H, Au L, Lee J, Li X, Xia Y (2005a) Gold nanocages: engineering their structure for biomedical applications. Adv Mater 17 (18):2255–2261

Chen J, Wiley B, Li ZY, Campbell D, Saeki F, Cang H, Xia Y (2005b) Gold nanocages: engineering their structure for biomedical applications. Adv Mate 17(18):2255–2261

Chen CT, Chen WJ, Liu CZ, Chang LY, Chen YC (2009) Glutathione-bound gold nanoclusters for selective-binding and detection of glutathione S-transferase-fusion proteins from cell lysates. Chem Commun 48:7515–7517

Chen ZY, Wang YX, Lin Y (2014) Advance of molecular imaging technology and targeted imaging agent in imaging and therapy. Biomed Res Int 2014:819324

Chen G, Roy I, Yang C, Prasad PN (2016a) Nanochemistry and nanomedicine for nanoparticle-based diagnostics and therapy. Chem Rev 116(5):2826–2885

Chen G, Roy I, Yang C, Prasad PN (2016b) Nanochemistry and nanomedicine for nanoparticle-based diagnostics and therapy. Chem Rev 116(5):2826–2885

Cheng Z, Dai Y, Kang X, Li C, Huang S, Lian H, Hou Z, Ma P, Lin J (2014a) Gelatin-encapsulated iron oxide nanoparticles for platinum (IV) prodrug delivery, enzyme-stimulated release and MRI. Biomaterials 35(24):6359–6368

Cheng H, Li Y, Huo K, Gao B, Xiong W (2014b) Long-lasting in vivo and in vitro antibacterial ability of nanostructured titania coating incorporated with silver nanoparticles. J Biomed Mater Res Part A 102(10):3488–3499

Ciccione J, Jia T, Coll JL, Parra K, Amblard M, Jebors S, Subra G (2016) Unambiguous and controlled one-pot synthesis of multifunctional silica nanoparticles. Chem Mater 28(3):885–889

Cormode DP, Skajaa T, Fayad ZA, Mulder WJ (2009) Nanotechnology in medical imaging: probe design and applications. Arter Thromb Vasc Biol 29(7):992–1000
Curry T, Kopelman R, Shilo M, Popovtzer R (2014) Multifunctional theranostic gold nanoparticles for targeted CT imaging and photothermal therapy. Contrast Media Mol Imaging 9(1):53–61
DaCosta MV, Doughan S, Han Y, Krull UJ (2014) Lanthanide upconversion nanoparticles and applications in bioassays and bioimaging: a review. Anal Chim Acta 832:1–33
Dolmans DE, Fukumura D, Jain RK (2003) Photodynamic therapy for cancer. Nat Rev Cancer 3 (5):380
Dong H, Du SR, Zheng XY, Lyu GM, Sun LD, Li LD, Yan CH (2015) Lanthanide nanoparticles: from design toward bioimaging and therapy. Chem Rev 115(19):10725–10815
Draz MS, Fang BA, Li L, Chen Z, Wang Y, Xu Y, Chen FF (2012) Hybrid nanocluster plasmonic resonator for immunological detection of hepatitis B virus. ACS Nano 6(9):7634–7643
Duncan R, Gaspar R (2011) Nanomedicine (s) under the microscope. Mol Pharm 8(6):2101–2141
El-Sayed IH, Huang X, El-Sayed MA (2006) Selective laser photo-thermal therapy of epithelial carcinoma using anti-EGFR antibody conjugated gold nanoparticles. Cancer Lett 239 (1):129–135
Geißler D, Charbonnière LJ, Ziessel RF, Butlin NG, Löhmannsröben HG, Hildebrandt N (2010) Quantum dot biosensors for ultrasensitive multiplexed diagnostics. Angew Chem Int Ed Engl 49 (8):1396–1401
Ghazani AA, Lee JA, Klostranec J, Xiang Q, Dacosta RS, Wilson BC, Chan WC (2006) High throughput quantification of protein expression of cancer antigens in tissue microarray using quantum dot nanocrystals. Nano Lett 6(12):2881–2886
Giepmans BN, Deerinck TJ, Smarr BL, Jones YZ, Ellisman MH (2005) Correlated light and electron microscopic imaging of multiple endogenous proteins using Quantum dots. Nat Methods 2(10):743
Giri S, Sykes EA, Jennings TL, Chan WC (2011) Rapid screening of genetic biomarkers of infectious agents using quantum dot barcodes. ACS Nano 5(3):1580–1587
Hainfeld JF, Slatkin DN, Smilowitz HM (2004) The use of gold nanoparticles to enhance radiotherapy in mice. Phys Med Biol 49(18):309–315
Hamd-Ghadareh S, Salimi A, Parsa S, Fathi F (2018) Simultaneous biosensing of CA125 and CA15-3 tumor markers and imaging of OVCAR-3 and MCF-7 cells lines via bi-color FRET phenomenon using dual blue-green luminescent carbon dots with single excitation wavelength. Int J Biol Macromol 118:617–628
He X, Bischof JC (2003) Quantification of temperature and injury response in thermal therapy and cryosurgery. Crit Rev Biomed Eng 31(5&6)
Heidegger S, Gößl D, Schmidt A, Niedermayer S, Argyo C, Endres S, Bourquin C (2016) Immune response to functionalized mesoporous silica nanoparticles for targeted drug delivery. Nanoscale 8(2):938–948
Hilderbrand SA, Shao F, Salthouse C, Mahmood U, Weissleder R (2009) Upconverting luminescent nanomaterials: application to in vivo bioimaging. Chem Commun 28:4188–4190
Hirsch LR, Stafford RJ, Bankson JA, Sershen SR, Rivera B, Price RE, Hazle JD, Halas NJ, West JL (2003) Nanoshell-mediated near-infrared thermal therapy of tumors under magnetic resonance guidance. PNAS 100(23):13549–13554
Hu S, Ye B, Yi X, Cao Z, Wu D, Shen C, Wang J (2016) Dumbbell-shaped metallothionein-templated silver nanoclusters with applications in cell imaging and Hg2+ sensing. Talanta 155:272–277
Huang X, El-Sayed IH, Qian W, El-Sayed MA (2006) Cancer cell imaging and photothermal therapy in the near-infrared region by using gold nanorods. J Am Chem Soc 128(6):2115–2120
Huang X, Jain PK, El-Sayed I, El-Sayed MA (2008) Plasmonic photothermal therapy (PPTT) using gold nanoparticles. Lasers Med Sci 23(3):217
Huang L, Dai T, Xuan Y, Tegos GP, Hamblin MR (2011a) Synergistic combination of chitosan acetate with nanoparticle silver as a topical antimicrobial: efficacy against bacterial burn infections. Antimicrob Agents Chemother 55(7):3432–3438

Huang X, Li L, Liu T, Hao N, Liu H, Chen D, Tang F (2011b) The shape effect of mesoporous silica nanoparticles on biodistribution, clearance, and biocompatibility in vivo. ACS Nano 5 (7):5390–5399

Huff TB, Tong L, Zhao Y, Hansen MN, Cheng JX, Wei A (2007) Hyperthermic effects of gold nanorods on tumor cells. Nanomed 2(1):125–132

Huh AJ, Kwon YJ (2011) "Nanoantibiotics": a new paradigm for treating infectious diseases using nanomaterials in the antibiotics resistant era. J Control Release 156(2):128–145

Jaffer FA, Libby P, Weissleder R (2007) Molecular imaging of cardiovascular disease. Circulation 116(9):1052–1061

Kamaly N, Xiao Z, Valencia PM, Radovic-Moreno AF, Farokhzad OC (2012) Targeted polymeric therapeutic nanoparticles: design, development and clinical translation. Chem Soc Rev 41 (7):2971–3010

Kiessling F, Mertens ME, Grimm J, Lammers T (2014) Nanoparticles for imaging: top or flop? Radiology 273(1):10–28

Kim TY, Kim DW, Chung JY, Shin SG, Kim SC, Heo DS, Kim NK, Bang YJ (2004) Phase I and pharmacokinetic study of Genexol-PM, a cremophor-free, polymeric micelle-formulated paclitaxel, in patients with advanced malignancies. Clin Cancer Res 10(11):3708–3716

Kim YP, Oh YH, Oh E, Ko S, Han MK, Kim HS (2008) Energy transfer-based multiplexed assay of proteases by using gold nanoparticle and quantum dot conjugates on a surface. Anal Chem 80 (12):4634–4641

Kobayashi H, Ogawa M, Alford R, Choyke PL, Urano Y (2009) New strategies for fluorescent probe design in medical diagnostic imaging. Chem Rev 110(5):2620–2640

Kumari A, Yadav SK, Yadav SC (2010) Biodegradable polymeric nanoparticles based drug delivery systems. Colloids Surf B Biointerfaces 75(1):1–18

Lammers T, Aime S, Hennink WE, Storm G, Kiessling F (2011) Theranostic nanomedicine. Acc Chem Res 44(10):1029–1038

Lammers T, Rizzo LY, Storm G, Kiessling F (2012) Personalized nanomedicine. Clin Cancer Res 18(18):4889–4894

Lee S, Cha EJ, Park K, Lee SY, Hong JK, Sun IC, Ahn CH (2008) A near-infrared-fluorescence-quenched gold-nanoparticle imaging probe for in vivo drug screening and protease activity determination. Angew Chem Int Ed 47(15):2804–2807

Lee CH, Cheng SH, Wang YJ, Chen YC, Chen NT, Souris J, Lo LW (2009) Near-infrared mesoporous silica nanoparticles for optical imaging: characterization and in vivo biodistribution. Adv Funct Mater 19(2):215–222

Li YY, Cunin F, Link JR, Gao T, Betts RE, Reiver SH, Chin V, Bhatia SN, Sailor MJ (2003) Polymer replicas of photonic porous silicon for sensing and drug delivery applications. Science 299(5615):2045–2047

Li N, Chang C, PanvW TB (2012a) A multicolor nanoprobe for detection and imaging of tumor-related mRNAs in living cells. Angew Chem 51(30):7426–7430

Li PH, Lin JY, Chen CT, Ciou WR, Chan PH, Luo L, Chen YC (2012b) Using gold nanoclusters as selective luminescent probes for phosphate-containing metabolites. Anal Chem 84 (13):5484–5488

Li L, Gao F, Ye J, Chen Z, Li Q, Gao W, Tang B (2013) FRET-based biofriendly apo-GO x-modified gold nanoprobe for specific and sensitive glucose sensing and cellular imaging. Anal Chem 85(20):9721–9727

Lim SF, Riehn R, Ryu WS, Khanarian N, Tung CK, Tank D, Austin RH (2006) In vivo and scanning electron microscopy imaging of upconverting nanophosphors in Caenorhabditis elegans. Nano Lett 6(2):169–174

Liong M, Lu J, Kovochich M, Xia T, Ruehm SG, Nel AE, Zink JI (2008) Multifunctional inorganic nanoparticles for imaging, targeting, and drug delivery. ACS Nano 2(5):889–896

Liu Z (2011) Nanoplatforms for Raman molecular imaging in biological systems. In: Nanoplatform-based molecular imaging. Wiley, Hoboken, pp 197–216

Liu J, Lau SK, Varma VA, Kairdolf BA, Nie S (2010) Multiplexed detection and characterization of rare tumor cells in Hodgkin's lymphoma with multicolor quantum dots. Anal Chem 82 (14):6237–6243

Liu JM, Chen JT, Yan XP (2013a) Near infrared fluorescent trypsin stabilized gold nanoclusters as surface plasmon enhanced energy transfer biosensor and in vivo cancer imaging bioprobe. Anal Chem 85(6):3238–3245

Liu D, Huang X, Wang Z, Jin A, Sun X, Zhu L, Chen X (2013b) Gold nanoparticle-based activatable probe for sensing ultralow levels of prostate-specific antigen. ACS Nano 7 (6):5568–5576

Liu X, Liu L, Hu X, Zhou S, Ankri R, Fixler D, Xie Z (2018) Multimodal bioimaging based on gold nanorod and carbon dot nanohybrids as a novel tool for atherosclerosis detection. Nano Res 11 (3):1262–1273

Loo C, Lin A, Hirsch L, Lee MH, Barton J, Halas N, West J, Drezek R (2004) Nanoshell-enabled photonics-based imaging and therapy of cancer. Technol Cancer Res Treat 3(1):33–40

Loo C, Lowery A, Halas N, West J, Drezek R (2005) Immunotargeted nanoshells for integrated cancer imaging and therapy. Nano Lett 5(4):709–711

Lu F, Wu SH, Hung Y, Mou CY (2009) Size effect on cell uptake in well-suspended, uniform mesoporous silica nanoparticles. Small 5(12):1408–1413

Lu W, Arumugam SR, Senapati D, Singh AK, Arbneshi T, Khan SA, Ray PC (2010) Multifunctional oval-shaped gold-nanoparticle-based selective detection of breast cancer cells using simple colorimetric and highly sensitive two-photon scattering assay. ACS Nano 4 (3):1739–1749

Maji SK, Sreejith S, Joseph J, Lin M, He T, Tong Y, Zhao Y (2014) Upconversion nanoparticles as a contrast agent for photoacoustic imaging in live mice. Adv Mater 26(32):5633–5638

Miao H, Wang L, Zhuo Y, Zhou Z, Yang X (2016) Label-free fluorimetric detection of CEA using carbon dots derived from tomato juice. Biosens Bioelectron 86:83–89

Michalet X, Pinaud FF, Bentolila LA, Tsay JM, Doose SJJL, Li JJ, Weiss S (2005) Quantum dots for live cells, in vivo imaging, and diagnostics. Science 307(5709):538–544

Minchin RF, Martin DJ (2010) Minireview: nanoparticles for molecular imaging—an overview. Endocrinology 151(2):474–481

Mintz K, Waidely E, Zhou Y, Peng Z, Al-Youbi AO, Bashammakh AS, El-Shahawi MS, Leblanc RM (2018) Carbon dots and gold nanoparticles based immunoassay for detection of alpha-L-fucosidase. Anal Chim Acta 1041:114–121

Moen MD, Lyseng-Williamson KA, Scott LJ (2009) Liposomal amphotericin B: a review of its use as empirical therapy in febrile neutropenia and in the treatment of invasive fungal infections. Drugs 69(3):361–392

Mohammadi S, Salimi A (2018) Fluorometric determination of microRNA-155 in cancer cells based on carbon dots and MnO 2 nanosheets as a donor-acceptor pair. Mikrochim Acta 185 (8):372

Mohammadi S, Salimi A, Hamd-Ghadareh S, Fathi F, Soleimani F (2018) A FRET immunosensor for sensitive detection of CA 15-3 tumor marker in human serum sample and breast cancer cells using antibody functionalized luminescent carbon-dots and AuNPs-dendrimer aptamer as donor-acceptor pair. Anal Biochem 557:18–26

Müller M, Müller-Zellenberg U, Hochhaus G, Derendorf H (2001) Current concepts in pharmacokinetics and their implications for clinical medicine. Wien Klin Wochenschr 113 (15–16):566–572

Nair LV, Philips DS, Jayasree RS, Ajayaghosh A (2013) A near-infrared fluorescent nanosensor (AuC@ Urease) for the selective detection of blood urea. Small 9(16):2673–2677

Nisman R, Dellaire G, Ren Y, Li R, Bazett-Jones DP (2004) Application of quantum dots as probes for correlative fluorescence, conventional, and energy-filtered transmission electron microscopy. J Histochem Cytochem 52(1):13–18

Noor MO, Krull UJ (2014) Camera-based ratiometric fluorescence transduction of nucleic acid hybridization with reagentless signal amplification on a paper-based platform using immobilized quantum dots as donors. Anal Chem 86(20):10331–10339

Ohulchanskyy TY, Roy I, Goswami LN, Chen Y, Bergey EJ, Pandey RK, Oseroff AR, Prasad PN (2007) Organically modified silica nanoparticles with covalently incorporated photosensitizer for photodynamic therapy of cancer. Nano Lett 7(9):2835–2842

Padmanabhan P, Kumar A, Kumar S, Chaudhary RK, Gulyas B (2016) Nanoparticles in practice for molecular-imaging applications: an overview. Acta Biomater 41:1–16

Pan L, Liu J, He Q, Wang L, Shi J (2013) Overcoming multidrug resistance of cancer cells by direct intranuclear drug delivery using TAT-conjugated mesoporous silica nanoparticles. Biomaterials 34(11):2719–2730

Paxton JW (1981) Elementary pharmacokinetics in clinical practice 3: practical pharmacokinetic applications. N Z Med J 94(696):381–384

Pelgrift RY, Friedman AJ (2013) Nanotechnology as a therapeutic tool to combat microbial resistance. Adv Drug Deliv Rev 65(13–14):1803–1815

Peng J, Feng LN, Zhang K, Li XH, Jiang LP, Zhu JJ (2012) Calcium carbonate–gold nanocluster hybrid spheres: synthesis and versatile application in immunoassays. Chem Eur J 18 (17):5261–5268

Petryayeva E, Algar WR (2015) Single-step bioassays in serum and whole blood with a smartphone, quantum dots and paper-in-PDMS chips. Analyst 140(12):4037–4045

Pirsaheb M, Mohammadi S, Salimi A (2019) Current advances of carbon dots based biosensors for tumor marker detection, cancer cells analysis and bioimaging. Trends Anal Chem 115(5):83–99

Prasad PN (2004) Introduction to biophotonics. Wiley, Hoboken

Prasad PN (2012) Introduction to nanomedicine and nanobioengineering, vol 7. Wiley, Hoboken

Probst CE, Zrazhevskiy P, Bagalkot V, Gao X (2013) Quantum dots as a platform for nanoparticle drug delivery vehicle design. Adv Drug Deliv Rev 65(5):703–718

Prodi L, Rampazzo E, Rastrelli F, Speghini A, Zaccheroni N (2015) Imaging agents based on lanthanide doped nanoparticles. Chem Soc Rev 44(14):4922–4952

Qiao J, Mu X, Qi L, Deng J, Mao L (2013) Folic acid-functionalized fluorescent gold nanoclusters with polymers as linkers for cancer cell imaging. Chem Commun 49(73):8030–8032

Qiu X, Hildebrandt N (2015) Rapid and multiplexed microRNA diagnostic assay using quantum dot based forster resonance energy transfer. ACS Nano 9(8):8449–8457

Radwan SH, Azzazy HM (2009) Gold nanoparticles for molecular diagnostics. Expert Rev Mol Diagn 9(5):511–524

Raj TA, Smith AM, Moore AS (2013) Vincristine sulfate liposomal injection for acute lymphoblastic leukemia. Int J Nanomedicine 8:4361–4369

Rizzo LY, Theek B, Storm G, Kiessling F, Lammers T (2013) Recent progress in nanomedicine: therapeutic, diagnostic and theranostic applications. Curr Opin Biotechnol 24(6):1159–1166

Roy I, Ohulchanskyy TY, Pudavar HE, Bergey EJ, Oseroff AR, Morgan J, Dougherty TJ, Prasad PN (2003) Ceramic-based nanoparticles entrapping water-insoluble photosensitizing anticancer drugs: a novel drug– carrier system for photodynamic therapy. J Am Chem Soc 125 (26):7860–7865

Schwartz-Cornil I, Mertens PP, Contreras V, Hemati B, Pascale F, Bréard E, Zientara S (2008) Bluetongue virus: virology, pathogenesis and immunity. Vet Res 39(5):46

Shang L, Azadfar N, Stockmar F, Send W, Trouillet V, Bruns M, Nienhaus GU (2011a) One-pot synthesis of near-infrared fluorescent gold clusters for cellular fluorescence lifetime imaging. Small 7(18):2614–2620

Shang L, Dörlich RM, Brandholt S, Schneider R, Trouillet V, Bruns M, Nienhaus GU (2011b) Facile preparation of water-soluble fluorescent gold nanoclusters for cellular imaging applications. Nanoscale 3(5):2009–2014

Shang L, Stockmar F, Azadfar N, Nienhaus GU (2013) Intracellular thermometry by using fluorescent gold nanoclusters. Angew Chem 52(42):11154–11157

Shi L, Hernandez B, Selke M (2006) Singlet oxygen generation from water-soluble quantum dot–organic dye nanocomposites. J Am Chem Soc 128(19):6278–6279

Song S, Liang Z, Zhang J, Wang L, Li G, Fan C (2009) Gold-nanoparticle-based multicolor nanobeacons for sequence-specific DNA analysis. Angew Chem 48(46):8670–8674

Takahashi H, Niidome T, Nariai A, Niidome Y, Yamada S (2006a) Gold nanorod-sensitized cell death: microscopic observation of single living cells irradiated by pulsed near-infrared laser light in the presence of gold nanorods. Chem Lett 35(5):500–501

Takahashi H, Niidome T, Nariai A, Niidome Y, Yamada S (2006b) Photothermal reshaping of gold nanorods prevents further cell death. Nanotechnology 17(17):4431

Theek B, Rizzo LY, Ehling J, Kiessling F, Lammers T (2014) The theranostic path to personalized nanomedicine. Clin Transl Imaging 2(1):67–76

Tokumasu F, Dvorak J (2003) Development and application of quantum dots for immunocytochemistry of human erythrocytes. J Microsc 211(3):256–261

Tsai TT, Shen SW, Cheng CM, Chen CF (2013) Paper-based tuberculosis diagnostic devices with colorimetric gold nanoparticles. Sci Technol Adv Mater 14(4):044404

Tsay JM, Trzoss M, Shi L, Kong X, Selke M, Jung ME, Weiss S (2007) Singlet oxygen production by peptide-coated quantum dot– photosensitizer conjugates. J Am Chem Soc 129 (21):6865–6871

Wang M, Mi CC, Wang WX, Liu CH, Wu YF, Xu ZR, Xu SK (2009) Immunolabeling and NIR-excited fluorescent imaging of HeLa cells by using NaYF4: Yb, Er upconversion nanoparticles. ACS Nano 3(6):1580–1586

Wang Y, Chen JT, Yan XP (2013) Fabrication of transferrin functionalized gold nanoclusters/graphene oxide nanocomposite for turn-on near-infrared fluorescent bioimaging of cancer cells and small animals. Anal Chem 85(4):2529–2535

Wang Y, Dai C, Yan XP (2014a) Fabrication of folate bioconjugated near-infrared fluorescent silver nanoclusters for targeted in vitro and in vivo bioimaging. Chem Commun 50(92):14341–14344

Wang LL, Qiao J, Liu HH, Hao J, Qi L, Zhou XP, Mao LQ (2014b) Ratiometric fluorescent probe based on gold nanoclusters and alizarin red-boronic acid for monitoring glucose in brain microdialysate. Anal Chem 86(19):9758–9764

Weissleder R, Pittet MJ (2008) Imaging in the era of molecular oncology. Nature 452(7187):580

Wolfbeis OS (2015) An overview of nanoparticles commonly used in fluorescent bioimaging. Chem Soc Rev 44(14):4743–4768

Xiao Y, Barker PE (2004) Semiconductor nanocrystal probes for human metaphase chromosomes. Nucleic Acids Res 32(3):28

Yezhelyev MV, Al-Hajj A, Morris C, Marcus AI, Liu T, Lewis M, Nie S (2007) In situ molecular profiling of breast cancer biomarkers with multicolor quantum dots. Adv Mater 19 (20):3146–3151

Zako T, Nagata H, Terada N, Utsumi A, Sakono M, Yohda M, Maeda M (2009) Cyclic RGD peptide-labeled upconversion nanophosphors for tumor cell-targeted imaging. Biochem Biophys Res 381(1):54–58

Zazo H, Millán CG, Colino CI, Lanao JM (2017) Applications of metallic nanoparticles in antimicrobial therapy. In: Antimicrobial nanoarchitectonics. Elsevier, Amsterdam, pp 411–444

Zhang H, Yee D, Wang C (2008) Quantum dots for cancer diagnosis and therapy: biological and clinical perspectives. Nanomed 3(1):83–91

Zhang Q, Wang X, Li PZ, Nguyen KT, Wang XJ, Luo Z, Zhang H, Tan NS, Zhao Y (2014) Biocompatible, uniform, and redispersible mesoporous silica nanoparticles for cancer-targeted drug delivery in vivo. Adv Funct Mater 24(17):2450–2461

Zharov VP, Galitovskaya EN, Johnson C, Kelly T (2005) Synergistic enhancement of selective nanophotothermolysis with gold nanoclusters: potential for cancer therapy. Lasers Surg Med 37 (3):219–226

Zheng M, Ruan S, Liu S, Sun T, Qu D, Zhao H, Xie Z, Gao H, Jing X, Sun Z (2015) Self-targeting fluorescent carbon dots for diagnosis of brain cancer cells. ACS Nano 9(11):11455–11461

Biocompatible Fluorescent Nanomaterials for Molecular Imaging Applications

3

Shanka Walia, Chandni Sharma, and Amitabha Acharya

Contents

Abstract

Fluorescence imaging by making use of promising fluorescent contrast agents allows noninvasive, sensitive, and real-time detection of biological events. In the past, imaging was done using conventional organic fluorophores, viz., organic dyes which suffer from photobleaching and spectral cross talk; cellular intolerance thus makes it difficult to use them for in vivo detection. Because of excellent

S. Walia · C. Sharma · A. Acharya (✉)
Biotechnology Division, CSIR-Institute of Himalayan Bioresource Technology, Palampur, Himachal Pradesh, India

Academy of Scientific & Innovative Research (AcSIR), CSIR-Institute of Himalayan Bioresource Technology, Palampur, Himachal Pradesh, India
e-mail: amitabha@ihbt.res.in

A. Acharya (ed.), *Nanomaterial - Based Biomedical Applications in Molecular Imaging, Diagnostics and Therapy*, https://doi.org/10.1007/978-981-15-4280-0_3

photoluminescence properties, robust chemical inertness, and low cytotoxicity, nanoparticles (NPs) are well explored for bioimaging applications over the past decade. So far safety concerns related to these NPs are still unaddressed, and thus NP application in clinical imaging is in the evaluation stage. However, in the future, these could be transformed into new diagnostic probes which can detect diseases at their early stage, thus improving both preclinical research and patient health.

Keywords

Fluorescence · Theranostic · Biocompatibility · Molecular imaging · Sensing

Abbreviations

%	Percentage
C-dot(s)	Carbon dots
CT	Computed tomography
MRI	Magnetic resonance imaging
NM(s)	Nanomaterials
NP(s)	Nanoparticles
PAMAM	Polyamidoamine
PAN-NPs	Polyacrylonitrile nanoparticles
PEG	Polyethylene glycol
PEI	Poly(ethylene imine)
PET	Positron emission tomography
PS-NPs	Polystyrene nanoparticles
QD(s)	Quantum dots
SERS	Surface-enhanced Raman spectroscopy
SiNP(s)	Si nanoparticles
SiQD(s)	Silicon quantum dots
SPECT	Single photon emission computed tomography
UCNPs	Upconversion nanoparticles
US	Ultrasound

3.1 Introduction

The development in nanotechnology has enthralled and got a lot of attention in the global society. Because of its application in the real world, this field has emerged as a multidisciplinary research which is aiming for understanding and manipulation of different materials at their molecular levels. This field is an outcome of different branches of sciences including chemistry, physics, engineering, medicine, biology, etc. (Buzea et al. 2007; Rathore and Sekhawat. 2011; Atsumi and Belcher 2018; Martínez-Ballesta et al. 2018). In recent times, the use of nanotechnology has become a need, because of its benchmark applications in agriculture, medicine, biology, electronics, agriculture, etc. The nanoparticles (NPs) because of their small size display some exceptional properties (viz., electrical, magnetic, optical, mechanical, and chemical), which are quite different from the bulk counterparts (Wetterskog et al. 2013; Talapin and Shevchenko 2016). Nanomaterials (NMs) have constantly been employed for the application in biomedical fields, viz., for biological imaging (Bogart et al. 2014; Zhou et al. 2014), as agents of drug delivery (Blanco et al. 2015; Gupta et al. 2018; Chakraborty et al. 2019), for applications in tissue engineering (Singla et al. 2019), as fluorescent probes and contrasting agents (Cho et al. 2013; Meder et al. 2016), and for in vitro as well as in vivo applications. The recent advancements in the nanotechnology-based diverse imaging modalities have clearly given immense importance to the NP-based biological imaging applications. Due to the unique NP characteristics, they have gained intense momentum in the clinical research and nanomedicine with their potential theranostic properties (Bruchez et al. 1998; Chan and Nie 1998; Giepmans et al. 2006; Choi and Wang 2011; Walia and Acharya 2016; Dar et al. 2019; Walia et al. 2019). Molecular imaging is of utmost importance for the early detection/diagnosis and treatment of any disease. Molecular imaging is a noninvasive and real-time technique used to identify, track, and monitor critical physiological and/or pathological processes occurring at cellular and molecular level in cells, tissues, and living organisms (Wu and Shu 2018; Han et al. 2019). In 2007, the Society of Nuclear Medicine and Molecular Imaging recommended definition for molecular imaging as "visualization, characterization, and measurement of biologic processes at the molecular and cellular levels" (Kim et al. 2017a, b, c). Molecular imaging enables the visualization of cellular and molecular functions by using highly sensitive and biocompatible contrast agent (Kim et al. 2017a, b, c). Each imaging modality provides important information which is required for molecular imaging. Clinically approved imaging modalities in current use include computed tomography (CT), X-ray radiography, magnetic resonance imaging (MRI), ultrasound (US), single photon emission computed tomography (SPECT), surface-enhanced Raman spectroscopy (SERS), positron emission tomography (PET), and fluorescence imaging (Padmanabhan et al. 2016; Walia and Acharya 2016; Li et al. 2016; Han et al. 2019). Among different imaging modalities, fluorescence imaging being noninvasive, rapid, highly sensitive, simple, and cost-effective played major role in clinical imaging (Giepmans et al. 2006). The exogenous contrasting agents can enhance the sensitivity or specificity of the data provided by any of the imaging modality (Key and Leary 2014). These contrast agents showed great potential in providing important diagnostic information; however, there are a few shortcomings, like they involve weak spatial and temporal resolution, nonspecific biodistribution, short blood circulation half-life, fast clearance, toxicity, and stability which challenge their clinical applications (Cormode et al. 2014; Kim et al. 2017a, b, c; Naseri et al. 2018).

To address these limitations, researchers are designing different nanoparticle-based contrast agents and reported groundbreaking results in molecular imaging. NP-based contrast agents using different elements, viz., Au, Ag, bismuth, tungsten, platinum, and europium for CT, X-ray, SERS, and photoacoustic imaging; Gd^{3+}, Mn^{2+}, and Fe^{3+} for MRI; quantum dots (QDs) for optical imaging; etc., have been successfully used for the diagnosis of severe diseases (Larson et al. 2003; Bharali et al. 2005; Yong et al. 2008; Kim et al. 2017a, b, c; Ehlerding et al. 2018; Pellico et al. 2019). Meanwhile NPs provide significant advantages over conventional imaging agents; along with this NPs also enable the combination of different imaging modalities, provide complete information, and also overcome the respective restrictions of the individual imaging modality (Li et al. 2016). This would offer more precise and inclusive physiological and anatomical data for diagnosis and treatment. In this regard fluorescent NPs are gaining utmost attention due to their high brightness/contrast and photostability (Kievit and Zhang 2011). This is because the fluorescent NPs are more resistant to photobleaching with high temporal stability and brightness than fluorescent proteins and conventional organic dyes (Pahari et al. 2018). In order to increase the clinical applications of these agents, some practical parameters, viz., safety, cellular tolerance, cost, and availability, need to be considered. NPs with unique property of large surface-to-volume ratio provide a suitable platform for surface modification and bioconjugation with increased sensitivity (Peng and Chiu 2015; Smith and Gambhir 2017). This enable the entry and targeted delivery of NP-based imaging agents into live cells without any toxic effect which is the basic need of biomedical applications. Fluorescence imaging due to its remarkable properties of high spatial resolution and high detection sensitivity can capture the shallow/deep images of cells and tissues (Achilefu 2010). In the past decade, this field has reached milestones, and researchers are expecting many more achievements in coming years. In this chapter, detailed discussions are made on the potential use of fluorescent NPs for in vitro and in vivo imaging applications. Finally, we included the challenges that fluorescent NPs face during biological imaging and also offered future perspectives on the respective field.

3.2 Fluorescence as a Basis of Molecular Imaging

The process of characterizing, visualizing, and measuring different biochemical processes at the cellular and molecular levels in the living systems such as human is known as molecular imaging. Molecular imaging is especially used for revealing the abnormalities in the biochemical processes at the cellular and molecular levels which are the potential causes for a particular disease (Weissleder and Mahmood 2001).In this process, the imaging as well as quantification of biological materials is done in two or three dimensions over a period of time (Mankoff 2007). There are a number of modalities that are used for molecular imaging. Each of which have their pros and cons, and some of them are multimodal in nature which make them more suitable for imaging at molecular levels with multiple targets than the other modes. These modalities are majorly classified as computed tomography (CT), magnetic resonance imaging (MRI), optical imaging, near-infrared imaging, single photon emission computed tomography (SPECT), positron emission tomography (PET), and ultrasound (US) (Willmann et al. 2008; Hadjipanayis et al. 2011; Vangestel et al. 2012; Peldschus and Ittrich 2014; Kajary and Molnár 2017; Kim et al. 2017a, b, c). The importance of fluorescence-based imaging in the present times has been at its exponentials because of its interesting and explorative properties. The technique of fluorescence has shown the potential of higher sensitivity, selectivity, contrasting properties, as well as versatile nature (Tainaka et al. 2010). From the last two decades, the field of fluorescence-based bioimaging has gained enormous attention because of its continuous refinement in the resolution and sensitivity that has even reached to the nanometer levels (Luo et al. 2006; Senesac and Thundat 2008; Li et al. 2013; Bai and Zhou 2014).The fluorescence-based imaging and sensing have been widely used in a number of application, viz., molecular and biomedical imaging, diagnostics, as well as analyte sensing (Walia and Acharya 2014; Dar et al. 2016; Walia and Acharya 2018). Although there are various advantages of the fluorescence-based molecular imaging, certain limitations are also associated with this technique. These include the decreased tissue penetration power of the fluorescence than many other techniques and also the presence of autofluorescence/scattering properties of the tissue under consideration. All these limitations hinder fluorescent materials to be used in the clinical and biomedical imaging (Hwang et al. 2006; Foucault-Collet et al. 2013).

3.3 Nanoparticles for Fluorescence Imaging Applications

Nanoparticles, because of their unique properties, can be used as a potential alternative to overcome the abovementioned limitations of fluorescence imaging. This can be achieved by using a number of fluorescent dyes loaded onto the NP surface which would help in providing much better signals than a NP with less number of dye molecules (Genovese et al. 2013). Additionally, surface modification or functionalization of NPs can be done to reduce or prevent the fluorescence quenching (Grebenik et al. 2013). Besides this, strategies (active/passive) can be used to increase the local concentration of NPs at the target sites, which ultimately will increase the sensitivity of the fluorescence imaging by increasing the concentration of fluorescent dyes present on the NP surface (Zhou et al. 2014). Keeping these potential advantages of fluorescent NMs under consideration, they have been used in molecular imaging, cell tracking, early-stage disease diagnosis, gene detection, tumor-related research, and many more as mentioned in Table 3.1 (Lee et al. 2008; Wei et al. 2012; Muthukumar et al. 2014; Wang et al. 2014; Markovic et al. 2016; Dubreil et al. 2017). Based on these applications, a number of fluorescent NMs have been used in different types of imaging and sensing applications; these include the fluorescent proteins, fluorescently doped silica NPs, quantum dots (QDs), upconversion NPs (UCNPs), noble metal NPs, silicon NPs (SiNPs), carbon dots (C-dots), hydrophilic/hydrophobic polymers, dendrimers/micelles, organic dyes, and many other NMs.

Table 3.1 Table summarizing different fluorescent nanoparticles and their corresponding bioimaging applications

S. no	NPs	Synthesis	Size (nm)	Quantum yield (%)	Emission (nm)	Application	References
1.	CdSe/CdZnS QDs	Non-injection heat up approach	8.7	–	680	Imaging and counting of mRNA in single cell	Liu et al. (2018)
2.	CdSe/ZnS QDs-IONP micelles	Refluxed method	50–70	4–20	590, 630	Long-term cell imaging and blood plasma protein binding studies	Chinnathambi et al. (2017)
3.	CdSe/CdS/ZnS QDs-SiO_2 NPs	Reverse microemulsion method/annealing/ photoactivation	30	56	575	Live cell imaging of HeLa cells	Wang et al. (2013a, b, c, d)
4.	Mesoporous SiO_2 NP-coated UCNPs ($NaErF_4$:Yb/Tm@$NaYF_4$: Yb@$NaNdF_4$:Yb)	Layer by layer deposition	20–28	–	540 and 650	In vitro imaging of A549 cells and in vivo real-time imaging-guided photodynamic therapy	Tang et al. (2019)
5.	Erythrocyte membrane-coated UCNPs		80–90	–	543 and 655	In vitro and in vivo imaging of MCF-7 human breast tumor	Rao et al. (2017)
6.	BSA/HSA-conjugated SiNPs	One pot synthesis	8.7	–	528	In vitro imaging of macrophage cells	Walia et al. (2017)
7.	BSA and antibody-conjugated SiQDs	Top-down method	130	40–45	621	In vitro imaging of live cancer cells	Tu et al. (2016)
8.	SiQDs	Optomechanical synchronization method	5	5.3	626	Ultrahigh-contrast time-gated in vivo fluorescence imaging using live zebrafish	Yang et al. (2019)
9.	SiNPs	One-step melt synthesis	5.7	13	502	Neuroblastoma cell imaging	Manhat et al. (2011)
10.	Methylene blue-loaded Au nanocluster-embedded mucin NPs	One-step green synthetic route	139 ± 47	3.5	580 and 700 nm	Photodynamic therapy and bioimaging of HeLa cells	Dutta et al. (2019)

11.	Pea plant protein-synthesized Au NPs	Green synthesis	100		635 nm	Bioimaging MCF-7 tumor xenograft-bearing mice	Li et al. (2018a, b)
12.	Cisplastin-conjugated Au NPs	Chemical synthesis	145 ± 22	2.16	590	In vitro imaging and therapy for HeLa cells	Chatterjee et al. (2018)
13.	AgNP-fluorescent glycine dimers	Thermal reduction	32	5.2	~480 nm	In vitro imaging of neural stem cells and rat basophilic leukemia cells	Kravets et al. (2016)
14.	N-doped C-dots	Hydrothermal synthesis	4.0	0.19	550–590	In vitro cell imaging and detection of Ag^+	Li et al. (2017a, b)
15.	C-dots	Maillard synthetic approach	3.3–5.8	72 and 6.5	520 and 596	Imaging and inhibition of bacterial and cancer cell progression	Walia et al. (2019)
16.	C-dots	Solvothermal	10	7.4	612	Single-molecule imaging of the nucleolus	Khan et al. (2018)
17.	Au-C-dots	Microwave-assisted method	2	9.3	470 and 565 nm	In vitro dual fluorescence imaging of breast cancer cells and normal rat osteoblast cells	Zhang et al. (2016a, b)
18.	Nitrogen- and chloride-doped C-dots	Chemical synthesis	3.88	16.15 ± 1.36	430	Biological and live cell imaging	Wang et al. (2018)

3.3.1 Fluorescently Doped Silica Nanoparticles

The first type of NPs that was used for molecular and bioimaging applications was silica NPs. Since then, a lot of research and their applications in biomedicine have been observed (Wang et al. 2013a, b, c, d). Wiesner and group described the use of silica NPs in different molecular imaging as well as sensing applications (Burns et al. 2006; Ma et al. 2012). Another form of silica, called the mesoporous silica NPs, has revolutionized the field of molecular imaging because of its remarkable properties. These NPs can be loaded with many fluorescent dyes, diagnostic molecules, etc. The mesoporous silica NPs showed less cytotoxicity with excellent molecular imaging capability (Zhou et al. 2014). These are generally doped with a number of luminescent lanthanides (Chen et al. 2014a, b; Zhang et al. 2014). These fluorophores impart many attractive features to the NPs like long fluorescence decay, narrow emission profiles, luminescence upconversion/down conversion, etc.

3.3.2 Quantum Dots

Semiconductor nanocrystals or quantum dots (QDs) exhibiting excitons in all three spatial dimensions were first discovered by Alexic Ekimov and Onushenko (1981). Since then the QDs were significantly used for in vitro and in vivo imaging applications. The first report on the use of QDs for biological imaging application came back in 1998 (Bruchez et al. 1998; Chan and Nie 1998). QDs are well known for their optical applications due to their remarkable property of high extinction coefficient (Matea et al. 2017).

QDs are particles with physical dimensions smaller than the exciton Bohr radius and particularly composed of groups IV (Si and Ge dots), II-VI (CdS, CdTe, CdSe or ZnS, ZnSe, etc.), III-V (InP), and I-III-VI ($CuInS_2$, $CuInSe_2$) elements (Alivisatos 1996). When such an element is hit by visible light, some of its electrons absorb a photon and thus are excited into higher-energy states (Chinnathambi and Shirahata 2019). On coming back to the ground state, a photon with a frequency corresponding to the characteristic material is emitted by these electrons. QDs prepared from metal and semiconductors mostly (2–6 nm) are used for biological applications because their dimensions resemble with most of the biological macromolecules such as nucleic acids, proteins, etc.

Before their discovery fluorescent organic dyes were used for biological applications which suffer from drawbacks of photobleaching. To this end, QDs filled the void and showed better results than the conventional dyes in terms of emission, multiple color imaging, photostability, and brightness (Abbasi et al. 2016). Khalid et al. explored the intrinsic fluorescence properties of selenium NPs for in vitro imaging and tracking of fibroblast cells without the use of additional fluorescent probes (Khalid et al. 2016). The NPs showed distinct fluorescence in the visible to NIR range centered at 580 nm. Glutathione-capped CdSe@ZnS QDs were prepared via facile ligand exchange method (Chen et al. 2019). These QDs showing maximum emission of 520 nm with size of 2.5 nm were used for plasma membrane labeling within 30 min and continuous membrane potential imaging for up to 60 min. Further, the surface modification of QDs with drugs or biomolecules such as proteins, peptides, antibodies, or polymers provided the benefits of targeted delivery, sensitive imaging, and diagnosis for both in vitro and in vivo applications (Rosenthal et al. 2011). Nowadays, NIR QDs with emission wavelength ranging from 700 nm to 900 nm are playing a major role in biological imaging applications. The main highlights of NIR QDs include high fluorescence intensity, high signal-to-noise ratios, excellent photostability and sensitivity, high spatial resolution, maximum penetration depth in tissues, and minimal tissue autofluorescence interference (Ren et al. 2017). The major drawback of these QDs is the toxicity associated with the heavy metals such as Cd, Pb, etc. which affects cells and organisms and thus hindered their applications in biological systems (Jin et al. 2011). Protecting the core of these bare QDs by using ligands or shells can reduce the leakage of toxic metal ions caused due to photolysis and oxidation into biological systems to some extent (Derfus et al. 2004; Guo et al. 2006). Brunetti et al. (2018) reported NIR polyethylene glycol (PEG)-coated CdSe/ZnS QDs functionalized with NT4 cancer-targeted tetra-branched peptide for in vitro and in vivo imaging. The results showed selective binding and higher retention of these QDs in human cancer cells and in colon cancer mouse model. Recently, Yan et al. (2019) reported optical imaging of inflamed tissues using $CuFeS_2$/ZnS core/shell QDs showing emission in the NIR region. The as-prepared QDs showed bright NIR fluorescence emission peak at 840 nm with quantum yield (QY) up to 52%. QDs were further coated with liposome and decorated with isolated macrophage membrane to target actively postoperative tumor-inflamed tissue. Reyes-Esparza et al. (2015) synthesized biocompatible dextrin-coated CdS NPs for in vitro and in vivo imaging applications. These NPs showed effective cellular uptake in HepG2 and HeLa cells, and bright fluorescence was observed in the lung and kidney followed by the liver, spleen, and brain of rodents. Biocompatible fatty acid-encapsulated NIR PbSe QDs were used for in vivo imaging of deep tissue (Shuhendler et al. 2011). Hydrophobic silica and amphiphilic polymer (Pluronic F-127)-functionalized NIR II emissive PbS@CdS QDs were synthesized via dual layer coating method. The water-dispersible QDs possess QY of ~5.79%. In vivo results showed that PbS@CdS@SiO_2@F-127 QDs were biocompatible and therefore utilized for real-time NIR II fluorescence imaging and deep tissue visualization (up to a depth of 950 μm) of blood vessels of the brain (Zebibula et al. 2018). Biocompatible fatty acid-encapsulated NIR PbSe QDs were used for in vivo imaging of deep tissue (Shuhendler et al. 2011). Fatty acid coating increased the QY of PbSe QDs up to 45%, and thus deep tissue fluorescence was recorded at emission wavelength of 840 nm on excitation of 710 nm. Also in vivo results showed nontoxic response, reduced liver uptake, increased tumor retention, and clearance of QDs from examined animals (Fig. 3.1).

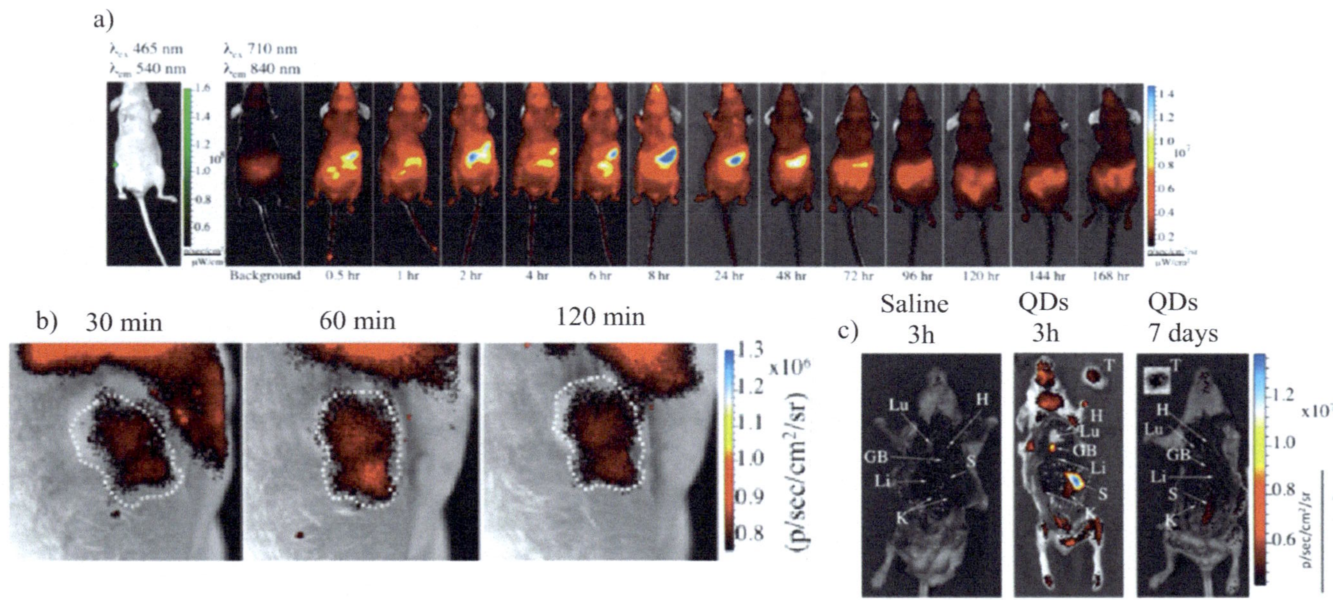

Fig. 3.1 (**a**) The biodistribution of fatty acid-coated PbSe QDs in a green fluorescent protein-expressing MDA435 model of human breast cancer was followed over 7 days with whole animal fluorescence imaging: (right) fluorescence image of the breast tumor; (left) longitudinal biodistribution and elimination of the QDs. (**b**) Tumor accumulation of the fatty acid–coated PbSe QDs was noted 30 min after intravenous administration. (**c**) Necropsy analysis of the tissue distribution showed significant accumulation of QDs in the spleen and gall bladder, without accumulation in the liver after 3 h. Near-complete elimination of the QDs was observed 7 days after administration: *GB* gall bladder, *H* heart, *K* kidney, *Li* liver, *Lu* lung, *S* spleen, *T* tumor. (Adapted from Shuhendler et al. (2011), reprint with permission)

3.3.3 Upconversion NPs

Upconversion NPs (UCNPs) are the most unique NMs that are doped with lanthanide ions giving them the property of increased electronic transitions. These are the NPs that convert two or more low-energy incident photons into one emitted photon of relatively higher energy (Liu et al. 2017; Wen et al. 2018). Over the years, the UCNPs achieved enormous attention because of their unique optical properties which enabled them to have many applications from biological sensing and light-mediated delivery of drugs to a more reliable super-resolution microscopy (Wilhelm 2017).

Most of the UCNPs are composed of $NaYF_4$ nanocrystals that are doped with different lanthanides such as Yb(III), Er(III), or Tm(III). As the trivalent lanthanide doping acts as the emitter, there is no need of further doping them with the fluorophores. UCNPs exhibit several colors of emission with the wavelengths that are dependent on the type of doped lanthanide. There is a great window for the tunability of the size of UCNPs which ranges from 10 nm to 100 nm with different quantum yields. The most difficult and challenging task for the usability of UCNPs is the control over their size and also the excitation/emission spectra (Sun et al. 2014). The fact that makes the UCNPs most suitable for the microscopy-based molecular imaging is that they display multi-photon emissions (Liu et al. 2014). In case of UCNPs, the color of the emission is independent upon the excitation wavelength. The UCNPs have been increasingly applied for bioimaging, drug delivery, therapy, and many bioassays (Yang 2014). Recently it was shown that silver coating on the UCNPs can cause plasmon-based upconversion with more than 30-fold increment in the brightness relative to the NPs (Yin et al. 2014). A number of UCNPs have been used for cellular- and molecular-based bioimaging as well as facilitating multiplexed imaging of cells (Lu et al. 2014).

The UCNPs are the perfect candidate molecules which are best suited for molecular bioimaging, because there is no interference of tissue/cell autofluorescence (Mader et al. 2010). Further, they also showed very good cytocompatibility with high tissue/cell permeability (Hemmer et al. 2014). The UCNPs usually possess a little interaction with the proteins, and hence no protein corona formation could hinder their biological applicability (Liebherr et al. 2012). UCNPs do not show any type of immune system activation under in vivo conditions. The UCNPs can be modified with ease; they also display no swelling but under ionic influence (generally bivalent ions) tend to aggregate quickly.

3.3.4 Noble Metal NPs

The salient features of the noble metal NPs (gold, silver, and platinum NPs) include their excellent photostability, solubility in water, no swelling in aqueous solutions, colors which are size dependent, excellent contrast, easy tunability, and characterization by different state-of-the-art spectroscopy and microscopy studies. Their surface functionalization is generally done for targeted imaging or biosensing applications. Dykman and Khlebtsov (2013) used gold NPs for imaging and their uptake by the mammalian cells. The gold and silver NPs were used for fluorescence imaging of HeLa cells (He et al. 2008). It has been reported that the metal clusters of gold and silver NPs display a strong intrinsic fluorescence (Lu and Chen 2012). The noble metal NPs being inert are highly biocompatible and thus preferred for their use in molecular imaging applications. With proper modification and change in the surface chemistry of the gold and silver NPs, they can be used for both specific bioimaging and therapeutic applications. The fluorescence is often credited to their particle size, due to which there occurs a size-dependent change in the fluorescent property of these NPs (Shang et al. 2011). The spectra and the fluorescence decay of noble metal NPs are mostly dependent on their structural parameters like the ligands present on their surface, valence state, and crystallinity. Various types of green, red, and yellow luminescent silver NPs have been used for bioimaging purpose (Díez and Ras 2011).

The noble metal NPs are known to provide excellent contrast in different types of imaging modalities like CT, fluorescence and SERS, etc. (Curry et al. 2014). Besides being good at molecular imaging, they are also potential candidates for the targeted drug deliveries by simple surface modifications with certain peptides or antibodies. These NPs are very selective and sensitive toward their imaging capabilities. The combinatorial process of gold NPs with other moieties has further enabled to use them as multimodal theranostic agents. These NPs have been used in successful in vivo studies and visualization of pancreatic cancers (Chen et al. 2014a, b). Similarly, the use of silver NPs in cancer theranostics is just beginning to be explored at their preclinical stages. The silver NPs are having inherent antibacterial and anticancer efficacy and are used as contrasting agents in many bioimaging modalities (Mukherjee et al. 2014). Till date, most of the studies that have been done on the use of silver NPs for molecular imaging applications have been done under in vitro conditions, but also a few studies have been done in vivo. Tan et al. (2016) used polyaniline, indocyanine green, and PEG-coated silver NPs for simultaneous fluorescence as well as photoacoustic imaging.

3.3.5 Silicon NPs

During the past decade, QDs (CdS, CdSe) with unique optical properties are extensively used in preclinical imaging studies. Despite of significant achievements made by QDs, their toxicity always remains a matter of concern for human health. Since most of the QDs contain toxic heavy metals, it resulted in leaching of metal ions in the biological system. Hence, further research was focused on exploring alternatives to these QDs offering safety profile such as noble metals (gold, silver, platinum, and palladium), but they also deal with the limitations of manufacturing cost and requirement of large amount of material (So et al. 2018; Azharuddin et al. 2019). The diverse applications of robust silicon NPs (SiNPs) in biomedical field attracted the interest of scientists worldwide (Ji et al. 2018; Kustov et al. 2018). Silicon has been emerged as desirable replacement due to its higher abundance, high quantum efficiency, size-tunable emission, cost-effectiveness, and lower toxicity than heavy metal-based QDs (Walia et al. 2017; Park et al. 2019). Li et al. (2019) reported ^{19}F MR and fluorescence imaging of cancer cells using blue-emitting fluorinated SiNPs. In vivo results suggested that these NPs exhibit enhanced ^{19}F/^{1}H MRI contrast in rats with non-small cell lung tumors. Singh et al. (2019) synthesized monodispersed citrate-coated SiNPs with average size of 2.4 ± 0.5 nm. The NPs were further covalently equipped with a NIR dye (IR800-CW) and a radiolabeled chelating agent (^{64}Cu-NOTA = 1,4,7-triazacyclononane-1,4,7-triacetic acid). In vitro studies showed that the dual-labeled SiNPs were cytocompatible up to 100 μg/ml concentration and showed rapid renal clearance from A549 tumor-bearing BALB/c male nude mice as observed through PET and fluorescence imaging at a concentration of 10 mg/ml. The nanosized quantum effect of these NPs resulted in enhanced fluorescence signals from tumor site.

3.3.6 Carbon Dots

Carbon dots (C-dots) are generally defined as the clusters of carbon atoms (sometimes fractions of oxygen and hydrogen also) with diameters of typically 2–8 nm. C-dots are the most fascinating types of molecules that are used for molecular imaging and sensing applications. These materials consist of NPs having discrete, quasi-spherical nature and most importantly excellent biocompatibility under all the in vitro and in vivo conditions (Baker and Baker 2010; Bottini and Mustelin 2007). These types of NMs exhibiting smaller size and cost-effectiveness are particularly important in such conditions where biocompatibility, size, and cost are the critical requirements (Peng and Travas-Sejdic 2009). The C-dots have almost negligible swelling index in aqueous solution, but sometimes they tend to aggregate. The C-dots can be manipulated for increasing its fluorescence without the need of any type of dopants or fluorescent labels at all. The excitation and emission spectra are also very wide (~UV-650 nm).

The C-dot-based nanosystems were applied, for pH-sensitive bioimaging of the living cells and tissues (Kong et al. 2012). The functionalization and surface engineering of C-dots for enhanced molecular imaging are very difficult even when compared to the QDs (Ding et al. 2013). The synthesis of C-dots has been done using different forms of organic and natural materials that are rich in carbon; one recent example is the synthesis of fluorescent C-dots where carbohydrates were used as carbon source. The prepared C-dots showed excellent photostability, emission profile, high-fluorescence quantum yield of 72.5 ± 4.5, and good cellular and hemocompatibility. These C-dots were used as an imaging probe for both bacterial and cancerous cells (Walia et al. 2019). Cost-effective, multicolor fluorescent C-dots with high photoluminescence property have also been synthesized using laser ablation of graphite. The fluorescence of C-dots is generally improved by chemical treatment or the passivation approach of the surface; water-dispersible C-dots with tunable fluorescence have also been synthesized. All these synthesized C-dots were subsequently used for molecular imaging applications (Zhang et al. 2013). Zhang et al. (2016a, b) reported conjugation of fluorescent C-dots with hydrophobic organic dye capable of aggregation-induced emission (AIE), namely, AIEgen. This strategy was applied to solve the problem of fluorescence quenching caused by aggregation of NPs. These C-dots showed large Stokes shift, biocompatibility, high emission efficiency with emission of 550 nm, and strong photobleaching resistance. The C-dots conjugated with organic dye were successfully used for in vitro and in vivo noninvasive long-term bioimaging up to 21 days (Fig. 3.2).

3.3.7 Hydrophilic Polymers

Different types of hydrophilic materials have been used for NP-based molecular imaging applications because of their unique optical properties, biocompatibility, easy fabrication, and great performance. The important ones include the NP-based hydrogels and also certain natural polymers like cellulose. These hydrophilic polymers have higher biocompatibility with wide range of applications in molecular imaging. The NP-based hydrogels (nanogels) are soft with higher water solubility (Wang et al. 2012a, b, c). These hydrogels are highly permeable to different types of ionic species and many other organic hydrophilic molecules like glucose and amino acids. Commonly used hydrophilic polymers include polyurethanes, poly-(hydroxyethyl meth acrylamide), polyacrylamide (PAA), and to some extent poly-ethylene glycols (Wang et al. 2013a, b, c, d). The emission profiles of the NPs present in the nanogels can be manipulated to nearly any of the wavelength ranging from 300 nm to 1000 nm, using various types of dopants (organic, metal-organic fluorophore) (Wang et al. 2013a, b, c, d; Chen et al. 2014a, b). Nanogels are the most important NP-based hydrogels, because of their extreme soft nature with very high swelling index and with water uptake capacity from 10% to 90%. These gels are highly permeable to a myriad of hydrophilic species and can be made fluorescently labeled with different fluorescent probes. Nanogels have been used for imaging in different pH values inside the body as well as different cells (Peng et al. 2010). The fluorescent-labeled polymer nanogels have gained a lot of attention because they have opened the path for temperature-based imaging and sensing of cells (Wang et al. 2013a, b, c, d).

3.3.8 Hydrophobic Organic Polymers

The hydrophobic organic polymers that are being used for the molecular imaging application include the polystyrene NPs (PS-NPs), polyacrylonitrile NPs (PAN-NPs), and to some extent fluorescent-labeled poly(vinyl butyral) (Lundqvist et al. 2008). Polystyrene NPs being highly hydrophobic are generally doped with different types of nonpolar fluorophores having emission profile ranging from near UV which can be extended to more than 1000 nm (Wang et al. 2019). Their fluorescence and the decay time are easily tunable. The PS-NPs exhibit good biocompatibility as well as cytocompatibility and cell permeability. If they are

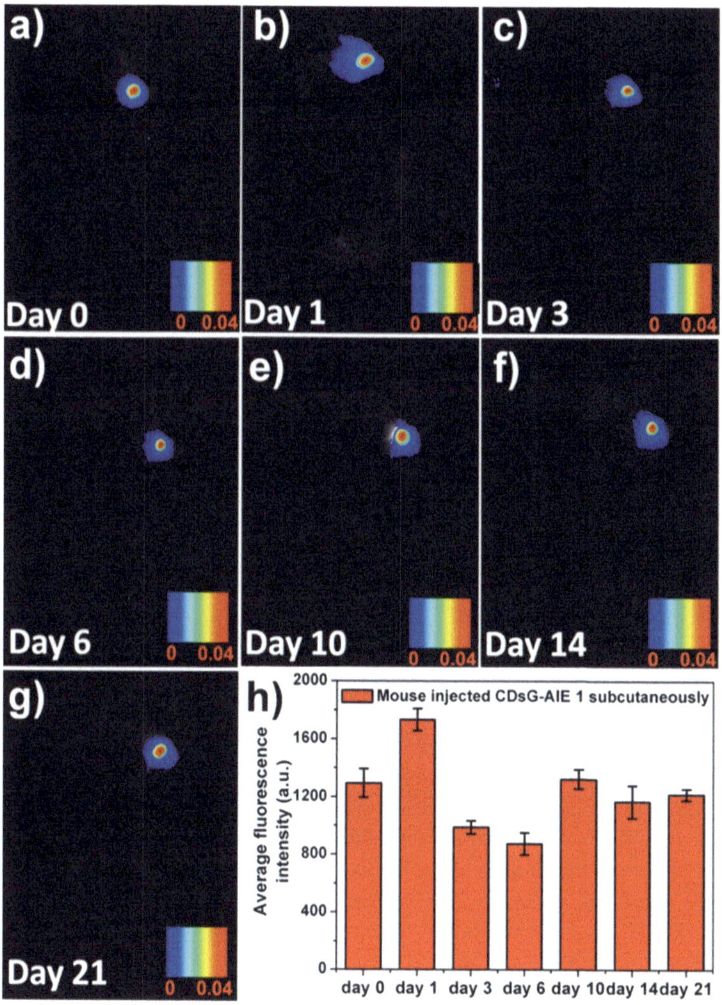

Fig. 3.2 (**a–g**) Representative time-dependent in vivo fluorescence images of the tumor-bearing mouse subcutaneously injected with CDsG-AIE 1 from day 0 to day 21. (**h**) Time-dependent fluorescence intensity changes for the tumors. Data represent mean values ± standard deviation, $n = 5$. (Adapted from Zhang et al. (2016), reprint with permission)

internalized inside the cells, they don't become heavily surrounded by the protein corona because they show very low immune system activation and recognition (Dar et al. 2019; Wang et al. 2019). The only barrier is that their surface modification is limited to few of the functional groups. Most important among these functionalities is the carboxyl and amino modification (Lunov et al. 2011; Loos et al. 2014). The first use of PS-NPs was for pH-based sensing, which consisted of the polyaniline-coated PS beads, in which the absorbance changes were observed by the change in pH values (Pringsheim et al. 2001).

PAN-NPs with the emission range of 300–1000 nm can be obtained by doping with different fluorophores; the most suitable among these modifications is by using the hydrophilic dyes. The fluorescence decay time and the size were widely tunable. PAN showed reasonable biocompatibility with the cells and tissues and is well permeable to the cells and shows no toxicity with low protein adsorption in biological milieu. Once modified, the surface of the PAN-NPs behaves inert and thus does not require any modification. The PAN-NPs doped with fluorophore were applied for pH- and temperature-sensitive imaging with a sub-micrometer spatial resolution (Wang et al. 2012a, b, c).

3.3.9 Semiconducting Polymer Dots

These organic polymer-based dots, also called the P-dots, are generally prepared from the polymerizable double or triple bonds containing aromatic precursors, either by the process of nanoprecipitation or by emulsion polymerization (Li and Liu 2012). The P-dots have strong fluorescence emission, which goes beyond the NIR. Like the most hydrophobic organic polymers, no such fluorophore doping is required in these NMs. The presence of conjugated polymers acts like the light-harvesting units, giving them the required larger optical cross section for imaging applications (Tang and Feng 2013). Moreover, these materials are generally not found to be photobleached. The possibility of wider tunability increases the fluorescent spectral properties as well as the targetability. These are very inert and don't show any type of swelling in water (Noh et al. 2014). P-dots show good biocompatibility, surface properties, and also colloidal dimensions. These are widely used for in vivo imaging of different types of cancers under hypoxia conditions, high-resolution imaging of subcellular structures, etc. (Yu et al. 2016; Zhou et al. 2016a, b). These semiconducting polymer fluorescent P-dots when attached with peptide like chlorotoxin can show excellent imaging even in the brain-related tumors (Wu et al. 2011).

3.3.10 Dendrimers, Lipid, and Micelles

Dendrimers are highly branched and spherical nanostructures that are produced synthetically. These are used as carrier species for different kinds of bioimaging applications. Different types of dendrimers exhibit versatile biological properties. Mostly, the core of the dendrimers is composed of molecules like diaminobutyl, ethylenediamine, and polyamidoamine (PAMAM). These dendrimers are also surface functionalized with different groups, viz., alcoholic, amine, and carboxyl moieties. Dendrimers being highly organized in their size and shape display small polydispersity index. Each of the dimensions of the dendrimer behaves differently having distinctive properties which provide many advantages for their medical applications (Malik et al. 2000). These NMs have been utilized for different types of imaging modalities including MRI, CT, optical imaging, as well as nuclear medicine (Wu et al. 1994). These have also been taken into consideration for drug delivery applications. Furthermore, due to the larger size and capacity for loading and functionalization in higher generation of dendrimers, they are often used for developing multimodal imaging, diagnostic, and delivery agents (Malik et al. 2000; Misselwitz et al. 2001). Dendrimers are easily internalized by cells because they are both hydrophilic and hydrophobic in nature. These lack inherent fluorescence but can be attached with fluorescent-labeled probe to its hyperbranched skeleton.

The fluorescence of dendrimer NPs greatly depends on its size and color of the fluorophore which can be manipulated easily. Because of the well establishment of the dendrimer chemistry, these have been increasingly applied for targeted imaging and sensing applications. Dendrimers possess very high molar absorbance and quantum yields. In a typical example, the dendrimer-based fluorescence imaging was reviewed in which these NPs were used to visualize pH changes in living HeLa cells (Albertazzi et al. 2010). Also a fifth-generation amino-terminated dendrimer was used for cell tracking by fluorescence microscopy (Gonçalves et al. 2014). Further, in addition to the labeled dendrimers, autofluorescent hyperbranched dendrimers made from PAMAM were used for cellular imaging (Yang et al. 2011). These dendrimers exhibit blue emission and high biocompatibility with no toxic effects on the cells.

Dendrimers are the outstanding candidates in the field of molecular imaging and diagnostics, but their biocompatibility, nanotoxicity, and degradation had not been discussed earlier. Dendrimers have been made more and more biocompatible by the surface modification like acetylation, glycosylation, and PEGylation and also by functionalizing them with amino acid. These modifications made these NMs very suitable for the therapeutic as well as diagnostic/imaging purposes. Further, a number of thrilling developments have been done on the dendrimer-based nanoplatforms for theranostic and bioimaging applications (Cheng et al. 2011).

3.3.11 Other Nanomaterials

There are a number of other NMs that have been employed for fluorescence-based molecular imaging applications in nanomedicine and nanotechnology. These NMs include phosphorous sulfides, metal oxides, fluorides, tellurides, and many other complexes of cationic and anionic species (Hui et al. 2012). Other typical examples of these NMs used for bioimaging are the LaF_3 NPs doped with Nd(III) and doping of fluoridated hydroxyapatites with the lanthanides like Eu(III), Tb(III), and Ln(III) (Zhang et al. 2013; Rocha et al. 2014). PEG-functionalized black phosphorus QDs were used for simultaneous bioimaging and combined photodynamic/photothermal cancer therapy in vitro and in vivo as well (Li et al. 2017a, b). The QDs prepared by sonication exfoliation exhibit emission at 577 nm and NIR photothermal and red-light-triggered photodynamic properties. All these NMs have a good biocompatibility as well as cellular imaging capability. The other NMs that have been used for molecular and bioimaging application are lanthanide- and phosphorus-doped YVO_4 (Wang et al. 2008), Eu(III)-doped titanium dioxide coated with PEI (Sandoval et al.

2012), glutathione-capped fluorescent Ag_2S (Wang et al. 2012a, b, c), Tm(III)- and Yb(III)-doped CaF_2 NPs (Dong et al. 2011), and some nanoscale metal-organic frameworks (Della et al. 2011).

3.4 Conclusions and Future Perspective

In this chapter, we have tried to summarize most of the developments made from time to time in fluorescent NP-based molecular and bioimaging applications and also discuss different biocompatibility issues for each type of NMs. We have discussed about various types of NMs including fluorescently doped silica NPs, QDs, UCNPs, noble metal NPs, C-dots, SiNPs, hydrophilic/hydrophobic polymers, semiconducting polymer dots, dendrimers/lipids and micelles, and many other NMs.

The modern state-of-the-art imaging based on fluorescence is very impressive and versatile. This versatility occurs due to the developments that are made in materials chemistry, nanotechnology, spectroscopy, and microscopy. A plethora of NMs have been used for fluorescence-based molecular imaging, due to which, the fluorescence imaging displays an intricacy and versatility in its nature. This creates a difficulty for choosing a perfect NP of interest for molecular imaging. Depending on the different types of application, various criteria have to be examined while selecting the materials. Nanoparticles having the size range close to most of the biomacromolecules are always preferred over the macroparticles for the intracellular as well as extracellular studies. For getting better molecular imaging applications using the NPs, certain criteria are needed to be taken into consideration for properly selecting them for their use in fluorescence-based imaging modality. These criteria are as follows:

1. Simple imaging needs biocompatibility and a good cell permeability. These include NPs like UCNPs, SiNPs, C-dots, biocompatible QDs, fluorescent-doped silica NPs or PS, and also the semiconductor P-dots.
2. The requirements for targeted cellular and bioimaging are high brightness, biocompatibility, and cell permeability, and their surface modification should be very easy like the silica NPs, silica-coated UCNPs, and also the C-dots.
3. The NPs that are to be used in chemical and analyte detection must be highly selective and chemical and temperature resistant with high biocompatibility with the biological and chemical environment.
4. NPs for temperature imaging should be entirely inert to other environmental changes, except dependence of their luminescence on the temperature.
5. NPs to be used for the multimodal imaging should be very sophisticated, versatile, and sensitive. These types of NPs display a sharp and bright fluorescence and also should possess multimodal imaging possibilities.
6. The most compatible NPs used for optical theranostics and imaging also require a prime sophistication, like the NPs that are having porous coating, in which any type of chemical moiety can be introduced.

Finally, it is a prime need of the time to comprehensively investigate the pros and cons of these NMs to be employed in molecular imaging applications, so that a generalization can be made for choosing the NMs with required features.

In the future, there is an intense need of collaboration between different researchers from diverse fields including physics, biology, chemistry, materials science, engineering, and pharmacy so that the NP-based imaging and theranostics can move forward. Also, more and more NP-based imaging demonstrations and simulations should be carried out under in vitro and in vivo conditions, because it will help in mimicking the visualization of cells, tissue, etc., which will provide a path for successful clinical trials even at the early stage. We hope that this chapter will give insights to the opportunities and challenges for further development in NP-based molecular and bioimaging agents followed by proper and safe clinical implementations.

Acknowledgments The authors would like to thank the director of CSIR-IHBT for his constant support and encouragement. AA acknowledges the financial support from CSIR (MLP-201) and DST (GAP-0214; EMR/2016/003027). SW acknowledges CSIR for the SRF fellowship, and CS thanks CSIR for providing JRF fellowship. The CSIR-IHBT communication number of this manuscript is 4536.

References

Abbasi E, Kafshdooz T, Bakhtiary M, Nikzamir N, Nikzamir N, Nikzamir M, Mohammadian M, Akbarzadeh A (2016) Biomedical and biological applications of quantum dots. Artif Cells Nanomed Biotechnol 44(3):885–891

Achilefu S (2010) Introduction to concepts and strategies for molecular imaging. Chem Rev 110:2575–2578

Albertazzi L, Storti B, Marchetti L, Beltram F (2010) Delivery and subcellular targeting of dendrimer-based fluorescent pH sensors in living cells. J Am Chem Soc 132(51):18158–18167

Alivisatos AP (1996) Semiconductor clusters, nanocrystals, and quantum dots. Science 271 (5251):933–937

Atsumi H, Belcher AM (2018) DNA origami and G-quadruplex hybrid complexes induce size control of single-walled carbon nanotubes via biological activation. ACS Nano 12 (8):7986–7995

Azharuddin M, Zhu GH, Das D, Ozgur E, Uzun L, Turner APF, Patra HK (2019) A repertoire of biomedical applications of noble metal nanoparticles. Chem Commun 55(49):6964–6996

Bai J, Zhou B (2014) Titanium dioxide nanomaterials for sensor applications. Chem Rev 114 (19):10131–10176

Baker SN, Baker GA (2010) Luminescent carbon nanodots: emergent nanolights. Angew Chem Int Ed Engl 49(38):6726–6744

Bharali DJ, Lucey DW, Jayakumar H, Pudavar HE, Prasad PN (2005) Folate-receptor-mediated delivery of InP quantum dots for bioimaging using confocal and two-photon microscopy. J Am Chem Soc 127(32):11364–11371

Blanco E, Shen H, Ferrari M (2015) Principles of nanoparticle design for overcoming biological barriers to drug delivery. Nat Biotechnol 33(9):941–951

Bogart LK, Pourroy G, Murphy CJ, Puntes V, Pellegrino T, Rosenblum D, Peer D, Lévy R (2014) Nanoparticles for imaging, sensing, and therapeutic intervention. ACS Nano 8(4):3107–3122

Bottini M, Mustelin T (2007) Carbon materials: nanosynthesis by candlelight. Nat Nanotechnol 2 (10):599–600

Bruchez M Jr, Moronne M, Gin P, Weiss S, Alivisatos AP (1998) Semiconductor nanocrystals as fluorescent biological labels. Science 281(5385):2013–2016

Brunetti J, Riolo G, Gentile M, Bernini A, Paccagnini E, Falciani C, Lozzi L, Scali S, Depau L, Pini A, Lupetti P (2018) Near-infrared quantum dots labelled with a tumor selective tetrabranched peptide for in vivo imaging. J Nanobiotechnol 16(1):21

Burns A, Sengupta P, Zedayko T, Baird B, Wiesner U (2006) Core/shell fluorescent silica nanoparticles for chemical sensing: towards single-particle laboratories. Small 2(6):723–726

Buzea C, Pacheco II, Robbie K (2007) Nano-materials and nanoparticles sources and toxicity. Biointerphases 2(4):MR17–MR71

Chakraborty D, Mohan L, Alex SA, Chandrasekaran N, Mukherjee A (2019) Bimetallic gold nanorods with enhanced biocorona formation for doxorubicin loading and sustained release. Biomater Sci 7(1):63–75

Chan WC, Nie S (1998) Quantum dot bioconjugates for ultrasensitive nonisotopic detection. Science 281(5385):2016–2018

Chatterjee B, Ghoshal A, Chattopadhyay A, Ghosh SS (2018) dGTP-templated luminescent gold nanocluster-based composite nanoparticles for cancer theranostics. ACS Biomater Sci Eng 4 (3):1005–1012

Chen W, Ayala-Orozco C, Biswal NC, Perez-Torres C, Bartels M, Bardhan R, Stinnet G, Liu XD, Ji B, Deorukhkar A, Brown LV, Guha S, Pautler RG, Krishnan S, Halas NJ, Joshi A (2014a) Targeting pancreatic cancer with magneto-fluorescent theranostic gold nanoshells. Nanomedicine 9(8):1209–1222

Chen X, Liu Y, Tu D (2014b) Bioimaging based on lanthanide-doped nanoprobes. In: Lanthanide-doped luminescent nanomaterials. Nanomedicine and nanotoxicology. Springer, Berlin, pp 145–164

Chen G, Zhang Y, Peng Z, Huang D, Li C, Wang Q (2019) Glutathione-capped quantum dots for plasma membrane labeling and membrane potential imaging. Nano Res 12:1–6

Cheng Y, Zhao L, Li Y, Xu T (2011) Design of biocompatible dendrimers for cancer diagnosis and therapy: current status and future perspectives. Chem Soc Rev 40(5):2673–2703

Chinnathambi S, Shirahata N (2019) Recent advances on fluorescent biomarkers of near-infrared quantum dots for in vitro and in vivo imaging. Sci Technol Adv Mater 20(1):337–355

Chinnathambi S, Abu N, Hanagata N (2017) Biocompatible CdSe/ZnS quantum dot micelles for long-term cell imaging without alteration to the native structure of the blood plasma protein human serum albumin. RSC Adv 7(5):2392–2402

Cho H, Alcantara D, Yuan H, Sheth RA, Chen HH, Huang P, Andersson SB, Sosnovik DE, Mahmood U, Josephson L (2013) Fluorochrome-functionalized nanoparticles for imaging DNA in biological systems. ACS Nano 7(3):2032–2041

Choi J, Wang NS (2011) Nanoparticles in biomedical applications and their safety concerns. In: Fazel-Rezai R (ed) Biomedical engineering-from theory to applications. InTechOpen, Rijeka, pp 209–314

Cormode DP, Naha PC, Fayad ZA (2014) Nanoparticle contrast agents for computed tomography: a focus on micelles. Contrast Media Mol Imaging 9(1):37–52

Curry T, Kopelman R, Shilo M, Popovtzer R (2014) Multifunctional theranostic gold nanoparticles for targeted CT imaging and photothermal therapy. Contrast Media Mol Imaging 9(1):53–61

Dar AI, Walia S, Acharya A (2016) Citric acid-coated gold nanoparticles for visual colorimetric recognition of pesticide dimethoate. J Nanopart Res 18(8):233

Dar AI, Walia S, Acharya A (2019) Molecular recognition based rapid diagnosis of immunoglobulins via proteomic profiling of protein-nanoparticle complexes. Int J Biol Macromol 138:156–167

Della Rocca J, Liu D, Lin W (2011) Nanoscale metal-organic frameworks for biomedical imaging and drug delivery. Acc Chem Res 44(10):957–968

Derfus AM, Chan WC, Bhatia SN (2004) Probing the cytotoxicity of semiconductor quantum dots. Nano Lett 4(1):11–18

Díez I, Ras RH (2011) Fluorescent silver nanoclusters. Nanoscale 3(5):1963–1970

Ding C, Zhu A, Tian Y (2013) Functional surface engineering of C-dots for fluorescent biosensing and in vivo bioimaging. Acc Chem Res 47(1):20–30

Dong NN, Pedroni M, Piccinelli F, Conti G, Sbarbati A, Ramírez-Hernández JE, Maestro LM, Iglesias-de la Cruz MC, Sanz-Rodriguez F, Juarranz A, Chen F, Vetrone F, Capobianco JA, Solé JG, Bettinelli M, Jaque D, Speghini A (2011) NIR-to-NIR two-photon excited CaF_2: Tm^{3+}, Yb^{3+} nanoparticles: multifunctional nanoprobes for highly penetrating fluorescence bio-imaging. ACS Nano 5(11):8665–8671

Dubreil L, Leroux I, Ledevin M, Schleder C, Lagalice L, Lovo C, Fleurisson R, Passemard S, Kilin V, Gerber-Lemaire S, Colle MA, Bonacina L, Rouger K (2017) Multi-harmonic imaging in the second near-infrared window of nanoparticle-labeled stem cells as a monitoring tool in tissue depth. ACS Nano 11(7):6672–6681

Dutta D, Sailapu SK, Simon AT, Ghosh SS, Chattopadhyay A (2019) Gold-nanocluster-embedded mucin nanoparticles for photodynamic therapy and bioimaging. Langmuir 35(32):10475–10483

Dykman LA, Khlebtsov NG (2013) Uptake of engineered gold nanoparticles into mammalian cells. Chem Rev 114(2):1258–1288

Ehlerding EB, Grodzinski P, Cai W, Liu CH (2018) Big potential from small agents: nanoparticles for imaging-based companion diagnostics. ACS Nano 12(3):2106–2121

Ekimov AI, Onuschenko AA (1981) The quantum size effect in three-dimensional microscopic semiconductors. JETP Lett 34(6):363

Foucault-Collet A, Gogick KA, White KA, Villette S, Pallier A, Collet G, Kieda C, Li T, Geib SJ, Rosi NL, Petoud S (2013) Lanthanide near infrared imaging in living cells with Yb^{3+} nano metal organic frameworks. Proc Natl Acad Sci 110(43):17199–17204

Genovese D, Bonacchi S, Juris R, Montalti M, Prodi L, Rampazzo E, Zaccheroni N (2013) Prevention of self-quenching in fluorescent silica nanoparticles by efficient energy transfer. Angew Chem Int Ed 52(23):5965–5968

Giepmans BN, Adams SR, Ellisman MH, Tsien RY (2006) The fluorescent toolbox for assessing protein location and function. Science 312(5771):217–224

Gonçalves M, Maciel D, Capelo D, Xiao S, Sun W, Shi X, Rodrigues J, Tomás H, Li Y (2014) Dendrimer-assisted formation of fluorescent nanogels for drug delivery and intracellular imaging. Biomacromolecules 15(2):492–499

Grebenik EA, Nadort A, Generalova AN, Nechaev AV, Sreenivasan VK, Khaydukov EV, Semchishen VA, Popov AP, Sokolov VI, Akhmanov AS, Zubov VP, Klinov DV, Panchenko VY, Deyev SM, Zvyagin AV (2013) Feasibility study of the optical imaging of a breast cancer lesion labeled with upconversion nanoparticle biocomplexes. J Biomed Opt 18(7):76004

Guo G, Liu W, Liang J, Xu H, He Z, Yang X (2006) Preparation and characterization of novel CdSe quantum dots modified with poly (D, L-lactide) nanoparticles. Mater Lett 60(21–22):2565–2568

Gupta A, Ahmad A, Dar AI, Khan R (2018) Synthetic lethality: from research to precision cancer nanomedicine. Curr Cancer Drug Targets 18(4):337–346

Hadjipanayis CG, Jiang H, Roberts DW, Yang L (2011) Current and future clinical applications for optical imaging of cancer: from intraoperative surgical guidance to cancer screening. Semin Oncol 38(1):109–118

Han X, Xu K, Taratula O, Farsad K (2019) Applications of nanoparticles in biomedical imaging. Nanoscale 11:799–819

He H, Xie C, Ren J (2008) Nonbleaching fluorescence of gold nanoparticles and its applications in cancer cell imaging. Anal Chem 80(15):5951–5957

Hemmer E, Vetrone F, Soga K (2014) Lanthanide-based nanostructures for optical bioimaging: small particles with large promise. Mater Res Soc Bull 39(11):960–964

Hui J, Zhang X, Zhang Z, Wang S, Tao L, Wei Y, Wang X (2012) Fluoridated HAp: Ln^{3+} (Ln= Eu or Tb) nanoparticles for cell-imaging. Nanoscale 4(22):6967–6970

Hwang ES, Cao C, Hong S, Jung HJ, Cha CY, Choi JB, Kim YJ, Baik S (2006) The DNA hybridization assay using single-walled carbon nanotubes as ultrasensitive, long-term optical labels. Nanotechnology 17(14):3442–3445

Ji X, Wang H, Song B, Chu B, He Y (2018) Silicon nanomaterials for biosensing and bioimaging analysis. Front Chem 6:38
Jin S, Hu Y, Gu Z, Liu L, Wu HC (2011) Application of quantum dots in biological imaging. J Nanomater 2011:13
Kajary K, Molnár Z (2017) The role of 18F-FDG PET/CT before and after the treatment of multiple myeloma: our clinical experience. Adv Mod Oncol Res 3(1):20–30
Key J, Leary JF (2014) Nanoparticles for multimodal in vivo imaging in nanomedicine. Int J Nanomedicine 9:711–726
Khalid A, Tran PA, Norello R, Simpson DA, O'Connor AJ, Tomljenovic-Hanic S (2016) Intrinsic fluorescence of selenium nanoparticles for cellular imaging applications. Nanoscale 8 (6):3376–3385
Khan S, Verma NC, Nandi CK (2018) Carbon dots for single-molecule imaging of the nucleolus. ACS Appl Nano Mater 1(2):483–487
Kievit FM, Zhang M (2011) Cancer nanotheranostics: improving imaging and therapy by targeted delivery across biological barriers. Adv Mater 23(36):H217–H247
Kim J, Chhour P, Hsu J, Litt HI, Ferrari VA, Popovtzer R, Cormode DP (2017a) Use of nanoparticle contrast agents for cell tracking with computed tomography. Bioconjug Chem 28(6):1581–1597
Kim J, Lee N, Hyeon T (2017b) Recent development of nanoparticles for molecular imaging. Philos Trans A Math Phys Eng Sci 375(2107):20170022
Kim T, Lee N, Arifin DR, Shats I, Janowski M, Walczak P, Hyeon T, Bulte JWM (2017c) In vivo micro-CT imaging of human mesenchymal stem cells labeled with gold-poly-L-lysine nanocomplexes. Adv Funct Mater 27(3):1604213
Kong B, Zhu A, Ding C et al (2012) Carbon dot-based inorganic-organic nanosystem for two-photon imaging and biosensing of pH variation in living cells and tissues. Adv Mater 24 (23):5844–5848
Kravets V, Almemar Z, Jiang K, Culhane K, Machado R, Hagen G, Kotko A, Dmytruk I, Spendier K, Pinchuk A (2016) Imaging of biological cells using luminescent silver nanoparticles. Nanoscale Res Lett 11(1):30
Kustov LM, Mashkin PV, Zakharov VN, Abramenko NB, Krysanov EY, Aslanov LA, Peijnenburg W (2018) Silicon nanoparticles: characterization and toxicity studies. Environ Sci Nano 5 (12):2945–2951
Larson DR, Zipfel WR, Williams RM, Clark SW, Bruchez MP, Wise FW, Webb WW (2003) Water-soluble quantum dots for multiphoton fluorescence imaging in vivo. Science 300:1434–1436
Lee S, Cha EJ, Park K, Lee SY, Hong JK, Sun IC, Kim SY, Choi K, Kwon IC, Kim K, Ahn CH (2008) A near-infrared-fluorescence-quenched gold-nanoparticle imaging probe for in vivo drug screening and protease activity determination. Angew Chem Int Ed 47(15):2804–2807
Li K, Liu B (2012) Polymer encapsulated conjugated polymer nanoparticles for fluorescence bioimaging. J Mater Chem 22(4):1257–1264
Li M, Gou H, Al-Ogaidi I, Wu N (2013) Nanostructured sensors for detection of heavy metals: a review. ACS Sustain Chem Eng 1:713–723
Li X, Zhang XN, Li XD (2016) Multimodality imaging in nanomedicine and nanotheranostics. Cancer Biol Med 13(3):339
Li J, Zuo G, Pan X, Wei W, Qi X, Su T, Dong W (2017a) Nitrogen-doped carbon dots as a fluorescent probe for the highly sensitive detection of Ag^+ and cell imaging. Luminescence 33 (1):243–248
Li Y, Liu Z, Hou Y, Yang G, Fei X, Zhao H, Guo Y, Su C, Wang Z, Zhong H, Zhuang Z, Guo Z (2017b) Multifunctional nanoplatform based on black phosphorus quantum dots for bioimaging and photodynamic/photothermal synergistic cancer therapy. ACS Appl Mater Interfaces 9 (30):25098–25106
Li S, Yuan Y, Yang Y, Li C, McMahon MT, Liu M, Chen S, Zhou X (2018a) Potential detection of cancer with fluorinated silicon nanoparticles in ^{19}F MR and fluorescence imaging. J Mater Chem B 6(26):4293–4300

Li Z, Peng H, Liu J, Tian Y, Yang W, Yao J, Shao Z, Chen X (2018b) Plant protein-directed synthesis of luminescent gold nanocluster hybrids for tumor imaging. ACS Appl Mater Interfaces 10(1):83–90

Liebherr RB, Soukka T, Wolfbeis OS, Gorris HH (2012) Maleimide activation of photon upconverting nanoparticles for bioconjugation. Nanotechnology 23(48):485103

Liu H, Xu CT, Andersson-Engels S (2014) Potential of multi-photon upconversion emissions for fluorescence diffuse optical imaging. Opt Express 22(15):17782–17790

Liu Y, Lu Y, Yang X, Zheng X, Wen S, Wang F, Vidal X, Zhao J, Liu D, Zhou Z, Ma C, Zhou J, Piper JA, Xi P, Jin D (2017) Amplified stimulated emission in upconversion nanoparticles for super-resolution nanoscopy. Nature 543(7644):229

Liu Y, Le P, Lim SJ, Ma L, Sarkar S, Han Z, Murphy SJ, Kosari F, Vasmatzis G, Cheville JC, Smith AM (2018) Enhanced mRNA FISH with compact quantum dots. Nat Commun 9(1):4461

Loos C, Syrovets T, Musyanovych A, Mailänder V, Landfester K, Simmet T (2014) Amino-functionalized nanoparticles as inhibitors of mTOR and inducers of cell cycle arrest in leukemia cells. Biomaterials 35(6):1944–1953

Lu Y, Chen W (2012) Sub-nanometre sized metal clusters: from synthetic challenges to the unique property discoveries. Chem Soc Rev 41(9):3594–3623

Lu D, Cho SK, Ahn S, Brun L, Summers CJ, Park W (2014) Plasmon enhancement mechanism for the upconversion processes in NaYF4:Yb($^{3+}$),Er($^{3+}$) nanoparticles: Maxwell versus Förster. ACS Nano 8(8):7780–7792

Lundqvist M, Stigler J, Elia G, Lynch I, Cedervall T, Dawson KA (2008) Nanoparticle size and surface properties determine the protein corona with possible implications for biological impacts. Proc Natl Acad Sci 105(38):14265–14270

Lunov O, Syrovets T, Loos C, Nienhaus GU, Mailänder V, Landfester K, Rouis M, Simmet T (2011) Amino-functionalized polystyrene nanoparticles activate the NLRP3 inflammasome in human macrophages. ACS Nano 5(12):9648–9657

Luo X, Morrin A, Killard AJ, Smyth MR (2006) Application of nanoparticles in electrochemical sensors and biosensors. Electroanalysis 18(4):319–326

Ma K, Sai H, Wiesner U (2012) Ultrasmall sub-10 nm near-infrared fluorescent mesoporous silica nanoparticles. J Am Chem Soc 4(32):13180–13183

Mader HS, Kele P, Saleh SM, Wolfbeis OS (2010) Upconverting luminescent nanoparticles for use in bioconjugation and bioimaging. Curr Opin Chem Biol 14(5):582–596

Malik N, Wiwattanapatapee R, Klopsch R (2000) Dendrimers: relationship between structure and biocompatibility in vitro, and preliminary studies on the biodistribution of 125I-labelled polyamidoamine dendrimers in vivo. J Control Release 65(1–2):133–148

Manhat BA, Brown AL, Black LA, Ross JB, Fichter K, Vu T, Richman E, Goforth AM (2011) One-step melt synthesis of water-soluble, photoluminescent, surface-oxidized silicon nanoparticles for cellular imaging applications. Chem Mater 23(9):2407–2418

Mankoff DA (2007) A definition of molecular imaging. J Nucl Med 48(6):18N–21N

Markovic S, Belz J, Kumar R, Cormack RA, Sridhar S, Niedre M (2016) Near-infrared fluorescence imaging platform for quantifying in vivo nanoparticle diffusion from drug loaded implants. Int J Nanomedicine 11:1213–1223

Martínez-Ballesta M, Gil-Izquierdo Á, García-Viguera C, Domínguez-Perles R (2018) Nanoparticles and controlled delivery for bioactive compounds: outlining challenges for new "smart-foods" for health. Foods 7(5):72

Matea CT, Mocan T, Tabaran F, Pop T, Mosteanu O, Puia C, Iancu C, Mocan L (2017) Quantum dots in imaging, drug delivery and sensor applications. Int J Nanomedicine 12:5421

Meder F, Thomas SS, Fitzpatrick LW, Alahmari A, Wang S, Beirne JG, Vaz G, Redmond G, Dawson KA (2016) Labeling the structural integrity of nanoparticles for advanced in situ tracking in bionanotechnology. ACS Nano 10(4):4660–4671

Misselwitz B, Schmitt-Willich H, Ebert W, Frenzel T, Weinmann HJ (2001) Pharmacokinetics of Gadomer-17, a new dendritic magnetic resonance contrast agent. MAGMA 12(2–3):128–134

Mukherjee S, Chowdhury D, Kotcherlakota R, Patra S, Vinothkumar B, Bhadra MP, Sreedhar B, Patra CR (2014) Potential theranostics application of bio-synthesized silver nanoparticles (4-in-1 system). Theranostics 4(3):316–335

Muthukumar T, Chamundeeswari M, Prabhavathi S, Gurunathan B, Chandhuru J, Sastry TP (2014) Carbon nanoparticle from a natural source fabricated for folate receptor targeting, imaging and drug delivery application in A549 lung cancer cells. Eur J Pharm Biopharm 88(3):730–736

Naseri N, Ajorlou E, Asghari F, Pilehvar-Soltanahmadi Y (2018) An update on nanoparticle-based contrast agents in medical imaging. Artif Cells Nanomed Biotechnol 46(6):1111–1121

Noh J, Kim D, Jang G, Kim J, Heo MB, Lee NE, Kim CY, Lee E, Kim YJ, Lim YT, Lee TS (2014) Fabrication, biofunctionalization, and simultaneous multicolor emission of hybrid "dots-on-spheres" structures for specific targeted imaging of cancer cells. RSC Adv 4(78):41378–41386

Padmanabhan P, Kumar A, Kumar S, Chaudhary RK, Gulyás B (2016) Nanoparticles in practice for molecular-imaging applications: an overview. Acta Biomater 41:1–16

Pahari SK, Olszakier S, Kahn I, Amirav L (2018) Magneto-fluorescent yolk-shell nanoparticles. Chem Mater 30(3):775–780

Park Y, Yoo J, Kang MH, Kwon W, Joo J (2019) Photoluminescent and biodegradable porous silicon nanoparticles for biomedical imaging. J Mater Chem B 7(41):6271–6292

Peldschus K, Ittrich H (2014) Magnetic resonance imaging of metastases in xenograft mouse models of cancer. In: Dwek M, Schumacher U, Brooks S (eds) Metastasis research protocols. Methods in molecular biology (Methods and protocols), vol 1070. Humana Press, New York, pp 213–222

Pellico J, Ellis CM, Davis JJ (2019) Nanoparticle-based paramagnetic contrast agents for magnetic resonance imaging. Contrast Media Mol Imaging 2019:1845637

Peng HS, Chiu DT (2015) Soft fluorescent nanomaterials for biological and biomedical imaging. Chem Soc Rev 44(14):4699–4722

Peng H, Travas-Sejdic J (2009) Simple aqueous solution route to luminescent carbogenic dots from carbohydrates. Chem Mater 21:5563–5565

Peng HS, Stolwijk JA, Sun LN, Wegener J, Wolfbeis OS (2010) A nanogel for ratiometric fluorescent sensing of intracellular pH values. Angew Chem Int Ed 49(25):4246–4249

Pringsheim E, Zimin D, Wolfbeis OS (2001) Fluorescent beads coated with polyaniline: a novel nanomaterial for optical sensing of pH. Adv Mater 13(11):819–822

Rao L, Meng QF, Bu LL, Cai B, Huang Q, Sun ZJ, Zhang WF, Li A, Guo SS, Liu W, Wang TH, Zhao XZ (2017) Erythrocyte membrane-coated upconversion nanoparticles with minimal protein adsorption for enhanced tumor imaging. ACS Appl Mater Interfaces 9(3):2159–2168

Rathore MS, Sekhawat NS (2011) A review on application of nanotechnology. Glob J Biotechnol Biochem 6:79–82

Ren D, Wang B, Hu C, You Z (2017) Quantum dot probes for cellular analysis. Anal Methods 9 (18):2621–2632

Reyes-Esparza J, Martínez-Mena A, Gutiérrez-Sancha I, Rodríguez-Fragoso P, de la Cruz GG, Mondragón R, Rodríguez-Fragoso L (2015) Synthesis, characterization and biocompatibility of cadmium sulfide nanoparticles capped with dextrin for in vivo and in vitro imaging application. J Nanobiotechnol 13:83

Rocha U, Kumar KU, Jacinto C, Villa I, Sanz-Rodríguez F, Iglesias de la Cruz Mdel C, Juarranz A, Carrasco E, van Veggel FC, Bovero E, Solé JG, Jaque D (2014) Neodymium-doped LaF_3 nanoparticles for fluorescence bioimaging in the second biological window. Small 10 (6):1141–1154

Rosenthal SJ, Chang JC, Kovtun O, McBride JR, Tomlinson ID (2011) Biocompatible quantum dots for biological applications. Chem Biol 18(1):10–24

Sandoval S, Yang J, Alfaro JG, Liberman A, Makale M, Chiang CE, Schuller IK, Kummel AC, Trogler WC (2012) Europium-doped TiO_2 hollow nanoshells: two-photon imaging of cell binding. Chem Mater 24(21):4222–4230

Senesac L, Thundat TG (2008) Nanosensors for trace explosive detection. Mater Today 11 (3):28–36

Shang L, Dong S, Nienhaus GU (2011) Ultra-small fluorescent metal nanoclusters: synthesis and biological applications. Nano Today 6(4):401–418

Shuhendler AJ, Prasad P, Chan HKC, Gordijo CR, Soroushian B, Kolios M, Yu K, O'Brien PJ, Rauth AM, Wu XY (2011) Hybrid quantum dot- fatty ester stealth nanoparticles: toward clinically relevant in vivo optical imaging of deep tissue. ACS Nano 5(3):1958–1966

Singh G, Ddungu JL, Licciardello N, Bermann R, De Cola L, Stephan H (2019) Ultrasmall silicon nanoparticles as promising platform for multimodal imaging. Faraday Discuss. https://doi.org/10.1039/C9FD00091G

Singla R, Abidi SMS, Dar AI, Acharya A (2019) Nanomaterials as potential and versatile platform for next generation tissue engineering applications. J Biomed Mater Res Part B 107 (7):2433–2449

Smith BR, Gambhir SS (2017) Nanomaterials for in vivo imaging. Chem Rev 117(3):901–986

So WY, Li Q, Legaspi CM, Redler B, Koe KM, Jin R, Peteanu LA (2018) Mechanism of ligand-controlled emission in silicon nanoparticles. ACS Nano 12(7):7232–7238

Sun LD, Wang YF, Yan CH (2014) Paradigms and challenges for bioapplication of rare earth upconversion luminescent nanoparticles: small size and tunable emission/excitation spectra. Acc Chem Res 47(4):1001–1009

Tainaka K, Sakaguchi R, Hayashi H, Nakano S, Liew FF, Morii T (2010) Design strategies of fluorescent biosensors based on biological macromolecular receptors. Sensors 10(2):1355–1376

Talapin DV, Shevchenko EV (2016) Introduction: nanoparticle chemistry. Chem Rev 116 (18):10343–10345

Tan X, Wang J, Pang X, Liu L, Sun Q, You Q, Tan F, Li N (2016) Indocyanine green-loaded silver nanoparticle@polyaniline core/shell theranostic nanocomposites for photoacoustic/nearinfrared fluorescence imaging-guided and single-light-triggered photothermal and photodynamic therapy. ACS Appl Mater Interfaces 8(51):34991–35003

Tang R, Feng X (2013) Highly luminescent conjugated polymer nanoparticles for imaging and therapy. Can Chem Trans 1:78–84

Tang M, Zhu X, Zhang Y, Zhang Z, Zhang Z, Mei Q, Zhang J, Wu M, Liu J, Zhang Y (2019) Near-infrared excited orthogonal emissive upconversion nanoparticles for imaging-guided on-demand therapy. ACS Nano 13(9):10405–10418

Tu CC, Chen KP, Yang TA, Chou MY, Lin LY, Li YK (2016) Silicon quantum dot nanoparticles with antifouling coatings for immunostaining on live cancer cells. ACS Appl Mater Interfaces 8 (22):13714–13723

Vangestel C, Van de Wiele C, Mees G, Mertens K, Staelens S, Reutelingsperger C, Pauwels P, Van Damme N, Peeters M (2012) Single-photon emission computed tomographic imaging of the early time course of therapy-induced cell death using technetium 99m tricarbonyl His-annexin A5 in a colorectal cancer xenograft model. Mol Imaging 11(2):7290–2011

Walia S, Acharya A (2014) Fluorescent cadmium sulfide nanoparticles for selective and sensitive detection of toxic pesticides in aqueous medium. J Nanopart Res 16(12):2778

Walia S, Acharya A (2016) Theragnosis: nanoparticles as a tool for simultaneous therapy and diagnosis. In: Yadav S (ed) Nanoscale materials in targeted drug delivery, theragnosis and tissue regeneration. Springer, Singapore, pp 127–152

Walia S, Acharya A (2018) Nanomaterials for chemical and biological analyte detection. J Bionanosci 12(3):316–327

Walia S, Guliani A, Acharya A (2017) A theragnosis probe based on BSA/HSA-conjugated biocompatible fluorescent silicon nanomaterials for simultaneous in vitro cholesterol effluxing and cellular imaging of macrophage cells. ACS Sustain Chem Eng 5(2):1425–1435

Walia S, Shukla AK, Sharma C, Acharya A (2019) Engineered bright blue-and red-emitting carbon dots facilitate synchronous imaging and inhibition of bacterial and cancer cell progression via $^1O_2$-mediated DNA damage under photoirradiation. ACS Biomater Sci Eng 5(4):1987–2000

Wang F, Xue X, Liu X (2008) Multicolor tuning of (Ln, P)-doped YVO_4 nanoparticles by single-wavelength excitation. Angew Chem Int Ed 47(5):906–909

Wang C, Wang Y, Xu L, Zhang D, Liu M, Li X, Sun H, Lin Q, Yang B (2012a) Facile aqueous-phase synthesis of biocompatible and fluorescent Ag_2S nanoclusters for bioimaging: tunable photoluminescence from red to near infrared. Small 8(20):3137–3142
Wang XD, Meier RJ, Wolfbeis OS (2012b) A fluorophore-doped polymer nanomaterial for referenced imaging of pH and temperature with sub-micrometer resolution. Adv Funct Mater 22(20):4202–4207
Wang XD, Stolwijk JA, Lang T, Sperber M, Meier RJ, Wegener J, Wolfbeis OS (2012c) Ultra-small, highly stable, and sensitive dual nanosensors for imaging intracellular oxygen and pH in cytosol. J Am Chem Soc 134(41):17011–17014
Wang K, He X, Yang X, Shi H (2013a) Functionalized silica nanoparticles: a platform for fluorescence imaging at the cell and small animal levels. Acc Chem Res 46(7):1367–1376
Wang X, Li W, Zhao B, Zhang D, Sun K, An X, Zhang Z, Shen Z (2013b) Highly fluorescent quantum dot@silica nanoparticles by a novel post-treatment for live cell imaging. RSC Adv 3 (11):3553–3556
Wang XD, Meier RJ, Wolfbeis OS (2013c) Fluorescent pH-sensitive nanoparticles in an agarose matrix for imaging of bacterial growth and metabolism. Angew Chem Int Ed 52(1):406–409
Wang XD, Wolfbeis OS, Meier RJ (2013d) Luminescent probes and sensors for temperature. Chem Soc Rev 42(19):7834–7869
Wang Y, Zhou K, Huang G, Hensley C, Huang X, Ma X, Zhao T, Sumer BD, DeBerardinis RJ, Gao J (2014) A nanoparticle-based strategy for the imaging of a broad range of tumours by nonlinear amplification of microenvironment signals. Nat Mater 13(2):204–212
Wang N, Zheng AQ, Liu X, Chen JJ, Yang T, Chen ML, Wang JH (2018) Deep eutectic solvent-assisted preparation of nitrogen/chloride-doped carbon dots for intracellular biological sensing and live cell imaging. ACS Appl Mater Interfaces 10(9):7901–7909
Wang H, Ma R, Nienhaus K, Nienhaus GU (2019) Formation of a monolayer protein corona around polystyrene nanoparticles and implications for nanoparticle agglomeration. Small 15 (22):1900974
Wei Y, Chen Q, Wu B, Zhou A, Xing D (2012) High-sensitivity in vivo imaging for tumors using a spectral up-conversion nanoparticle $NaYF_4$: Yb^{3+}, Er^{3+} in cooperation with a microtubulin inhibitor. Nanoscale 4(13):3901–3909
Weissleder R, Mahmood U (2001) Molecular imaging. Radiology 219(2):316–333
Wen S, Zhou J, Zheng K, Bednarkiewicz A, Liu X, Jin D (2018) Advances in highly doped upconversion nanoparticles. Nat Commun 9(1):1–12
Wetterskog E, Tai CW, Grins J, Bergström L, Salazar-Alvarez G (2013) Anomalous magnetic properties of nanoparticles arising from defect structures: topotaxial oxidation of Fe_{1-x} O|$Fe_{3-\delta}O_4$ core|shell nanocubes to single-phase particles. ACS Nano 7(8):7132–7144
Wilhelm S (2017) Perspectives for upconverting nanoparticles. ACS Nano 11(11):10644–10653
Willmann JK, Lutz AM, Paulmurugan R, Patel MR, Chu P, Rosenberg J, Gambhir SS (2008) Dual-targeted contrast agent for US assessment of tumor angiogenesis in vivo. Radiology 248 (3):936–944
Wu M, Shu J (2018) Multimodal molecular imaging: current status and future directions. Contrast Media Mol Imaging 2018:12
Wu CC, Brechbiel MW, Kozak RW, Gansow OA (1994) Metal-chelate-dendrimer-antibody constructs for use in radioimmunotherapy and imaging. Bioorg Med Chem 4(3):449–454
Wu C, Hansen SJ, Hou Q, Yu J, Zeigler M, Jin Y, Burnham DR, McNeill JD, Olson JM, Chiu DT (2011) Design of highly emissive polymer dot bioconjugates for in vivo tumor targeting. Angew Chem Int Ed 50(15):3430–3434
Yan T, Li Y, Song X, Wang J, Xie Z, Deng D (2019) Highly luminescent NIR-emitting CuFeS 2/ZnS core/shell quantum dots for optical imaging of inflamed tissue. J Mater Chem C 7:7279–7287
Yang Y (2014) Upconversion nanophosphors for use in bioimaging, therapy, drug delivery and bioassays. Microchim Acta 181(3–4):263–294

Yang W, Pan CY, Liu XQ, Wang J (2011) Multiple functional hyperbranched poly (amido amine) nanoparticles: synthesis and application in cell imaging. Biomacromolecules 12(5):1523–1531

Yang W, Srivastava PK, Han S, Jing L, Tu CC, Chen SL (2019) Optomechanical time-gated fluorescence imaging using long-lived silicon quantum dot nanoparticles. Anal Chem 91 (9):5499–5503

Yin D, Wang C, Ouyang J, Zhang X, Jiao Z, Feng Y, Song K, Liu B, Cao X, Zhang L, Han Y, Wu M (2014) Synthesis of a novel core-shell nanocomposite Ag@SiO_2@$Lu2O_3$: Gd/Yb/Er for large enhancing upconversion luminescence and bioimaging. ACS Appl Mater Interfaces 6 (21):18480–18488

Yong KT, Roy I, Pudavar HE, Bergey EJ, Tramposch KM, Swihart MT, Prasad PN (2008) Multiplex imaging of pancreatic cancer cells by using functionalized quantum rods. Adv Mater 20:1412–1417

Yu J, Rong Y, Kuo CT, Zhou XH, Chiu DT (2016) Recent advances in the development of highly luminescent semiconducting polymer dots and nanoparticles for biological imaging and medicine. Anal Chem 89(1):42–56

Zebibula A, Alifu N, Xia L, Sun C, Yu X, Xue D, Liu L, Li G, Qian J (2018) Ultrastable and biocompatible NIR-II quantum dots for functional bioimaging. Adv Funct Mater 28(9):1703451

Zhang X, Hui J, Yang B, Yang Y, Fan D, Liu M, Tao L, Wei Y (2013) PEGylation of fluoridated hydroxyapatite (FAp): Ln^{3+} nanorods for cell imaging. Polym Chem 4(15):4120–4125

Zhang Y, Wei W, Das GK, Tan TTY (2014) Engineering lanthanide-based materials for nanomedicine. J Photochem Photobiol C 20:71–96

Zhang J, Zheng M, Zhang F, Xu B, Tian W, Xie Z (2016a) Supramolecular hybrids of AIEgen with carbon dots for noninvasive long-term bioimaging. Chem Mater 28(23):8825–8833

Zhang L, Wang D, Huang H, Liu L, Zhou Y, Xia X, Deng K, Liu X (2016b) Preparation of gold-carbon dots and ratiometric fluorescence cellular imaging. ACS Appl Mater Interfaces 8 (10):6646–6655

Zhou H, Lv X, Zhang L, Gong A, Wu A, Liang Z, Peng G, Lin H (2014) A facile and in situ approach to fluorescent mesoporous silica and its applications in sensing and bioimaging. J Mater Chem C 2(45):9625–9630

Zhou M, Tian M, Li C (2016a) Copper-based nanomaterials for cancer imaging and therapy. Bioconjug Chem 27(5):1188–1199

Zhou X, Liang H, Jiang P, Zhang KY, Liu S, Yang T, Zhao Q, Yang L, Lv W, Yu Q, Huang W (2016b) Multifunctional phosphorescent conjugated polymer dots for hypoxia imaging and photodynamic therapy of cancer cells. Adv Sci 3(2):1500155

Nanomaterials for Point of Care Disease Detection

4

Chandni Sharma, Shanka Walia, and Amitabha Acharya

Contents

Abstract

In the recent era, nanomaterials (NMs) have been found useful for sensing as well as detection of several biomarkers, and these strategies are likely to develop point of care (PoC) diagnostics, which can be used for early disease detection and various therapeutic applications. Application of NMs in biosensing also allows the use of many novel signal transduction technologies in their preparation. The use of high surface area of NMs plays a crucial role in developing nanobiosensors. These nanoscale structures may include nanoparticles (NPs), nanotubes, nanorods, nanofibers, and thin films. Nanobiosensors have greater sensitivity, selectivity, and fast response time. Biosensing technologies include various types of techniques, viz., electrochemical, amperometric, voltametric,

C. Sharma · S. Walia · A. Acharya (✉)
Biotechnology Division, CSIR-Institute of Himalayan Bioresource Technology, Palampur, Himachal Pradesh, India

Academy of Scientific & Innovative Research (AcSIR), CSIR- Institute of Himalayan Bioresource Technology, Palampur, Himachal Pradesh, India
e-mail: amitabha@ihbt.res.in

A. Acharya (ed.), *Nanomaterial - Based Biomedical Applications in Molecular Imaging, Diagnostics and Therapy*, https://doi.org/10.1007/978-981-15-4280-0_4

fluorescence, etc. Particularly, fluorescence-based sensing technologies possess many advantages in their detection, viz., simple instrumentation, less background signals or noise, and having potential for multiplexed analysis. Similarly, colorimetric sensors have been well-known for their naked eye visualization, simplicity, and cost-effectiveness features. This chapter describes the advancements in promising NMs which are aimed for on-site clinical applications and those which can be used as a personalized diagnostic device in different disease conditions.

Keywords

Nanomaterials · Biosensing · Biomarkers · Point of care detection · Diagnosis

Abbreviations

AAB	Anti-Apolipoprotein B
Ab_1	Primary antibody
Ab_2	Target antibody
AD	Alzheimer's disease
ADDLs	Amyloid β-derived diffusible ligands
AgNPs	Silver nanoparticles
ApoE	Apolipoprotein E
AuNPs	Gold nanoparticles
BDBA	Benzenediboronic acid
BOx	Bilirubin oxidase
BSA	Bovine serum albumin
CdS	Cadmium sulfide
CDs	Carbon dots
CdSe@ZnS	Cadmium selenide/zinc sulfide
CEA	Carcinoembryonic antigen
ChE	Cholesterol esterase
ChOx	Cholesterol oxidase
CNPs	Carbon nanoparticles
CS@GIANs	Graphene-isolated-Au-nanocrystal-loaded cellulose paper strips
CVDs	Cardiovascular diseases
ECG	Electrocardiography
EGO	Exfoliated graphene oxide
EIS	Electrochemical impedance spectroscopy
FA	Folic acid
fap-SiNPs	Ferrocenyl iminopropyl-modified silica nanoparticle conjugates
FRET	Fluorescence resonance energy transfer
FTO	Fluorine-doped tin oxide
GCE	Glassy carbon electrode
GNWs	Gold nanowires
GONPs	Graphene oxide NPs
GOx	Glucose oxidase

GQDs	Graphene quantum dots
gQDs	Green-emitting quantum dots
HDL	High-density lipoprotein
IUPAC	International Union of Pure and Applied Chemistry
LDL	Low-density lipoprotein
LOD	Limit of detection
LOQ	Limit of quantification
LSPR	Localized surface plasmon resonance
MBs	Magnetic beads
miRNA	Micro-RNA
MUA	11-Mercaptoundecanoic acid
MWCNT	Multiwalled carbon nanotubes
NiHCFNPs	Nickel hexacyanoferrate NPs
NiO	Nickel oxide
NMs	Nanomaterials
NPG	Nanoporous gold
NPs	Nanoparticles
PCR	Polymerase chain reaction
PEDOT:PSS	Poly(3,4-ethylenedioxythiophene)-poly(styrene-sulfonate)
PoC	Point of care
PoCT	Point of care technologies
Ppy	Polypyrrole
PSA	Prostate-specific antigen
QDs	Quantum dots
rQDs	Red-emitting quantum dots
SERS	Surface-enhanced Raman scattering
SPEET	Surface plasmon-enhanced energy transfer
SPEs	Screen-printed carbon electrodes
SPR	Surface plasmon resonance
TC	Total cholesterol
TG	Triglyceride
TMB	3,3′,5,5′-Tetramethylbenzidine
WHO	World Health Organization
β-Gal	β-galactosidase

4.1 Introduction

Recent advancements in nanotechnology and its use in various fields have gained much interests, and this has become a topmost research area in the world. A NM consists of nanoparticles (NPs) which are less than 100 nm at least in one dimension (Jeevanandam et al. 2018; Cassano et al. 2017). Nanotechnology is having major impact on the development of an innovative class of biosensor known as nanobiosensors (Soleymani and Li 2017). The controlled synthesis and tuning properties of NMs require complete knowledge and awareness of different areas or disciplines such as chemistry, physics, biology, engineering, computer science, electronics, etc. which can lead to the development and designing of innovative and multifunctional technologies. The term "biosensor" was coined by Cammann (1977). Its definition was introduced by the International Union of Pure and Applied Chemistry (IUPAC), which states "biosensors are integrated receptor-transducer systems which are used to provide specific quantitative and qualitative information *via* biological recognition element" (Cammann 1977; Thévenot et al. 2001). Biosensors are one of the most important analytical tools which are responsible for converting biological response into an electrical signal (Rocchitta et al. 2016). Essentially, biosensors should be highly specific and reusable, and their characteristic properties should not vary in physiological pH and temperature conditions (Mehrotra 2016). Nanobiosensing system generally constitutes of a biologically recognizing moiety which is adsorbed or immobilized on the surface of a signaling transducer (Cui 2017). The interaction between the transducer and the analyte moiety is a heterogeneous type of reaction, and hence establishing the intermediate interface for biosensing is the most important element in evaluating the overall performance of the sensing system (Prasad 2014). In this context, the exciting properties of NMs have attracted much attention for their use in different fields, viz., food, health, transport, security, information technology sectors, etc. (Bajpai et al. 2018; Chung et al. 2017). The smart utilization of NMs is predicted to improve the performance of biomolecular electronic diagnostic devices with elevated sensitivity and low detection limits (Rizwan et al. 2018). Point of care technologies (PoCT) are novel detection techniques which assure to improve patient care. A PoCT development plan promised to direct multidisciplinary groups of scientists, technologists, healthcare professionals, and clinical trialists, as follows: (i) prepare demands and assessments, (ii) explain design qualifications of equipment, (iii) develop techniques and incorporate systems, (iv) execute repeatable bulk-scale evaluation, and (v) perform thorough potential clinical testing to make sure that PoCT solutions have significant effects (King Kevin et al. 2016). The schematic representation of various components and techniques has been illustrated in Fig. 4.1. Biosensors, due to being economical and user friendly (Freeman et al. 2013), have proven as a smart tool for routine testing of risk factors in home settings or primary care. It has been found that the sensitivity for measuring the lipid and lipoprotein level has increased significantly by integrating nanotechnology because of distinctive physicochemical properties of NMs (Lu et al. 2017). PoC diagnostics help to provide a quick and simple diagnosis in non-laboratory-based primary home settings which offer huge benefits over laboratory-based assays for disease prevention and monitoring the recovery (Nayak et al. 2016). PoC analytical platforms, viz., microfluidic devices, lateral flow assays (LFA), and biosensing tools, have been developed for the detection of various infectious diseases (Rozand 2014) and physiological status like pregnancy (Balakrishnan et al. 2015), monitoring the glucose level in blood samples, etc. (Christodouleas et al. 2018; Ding and Yang 2013). The major goal of PoCT is to provide a simple, small, chip-based, portable system which allows the quantification of various biomarkers like nucleic acids, proteins, and cells in composite samples (Sun et al. 2014). Biosensors constitute both

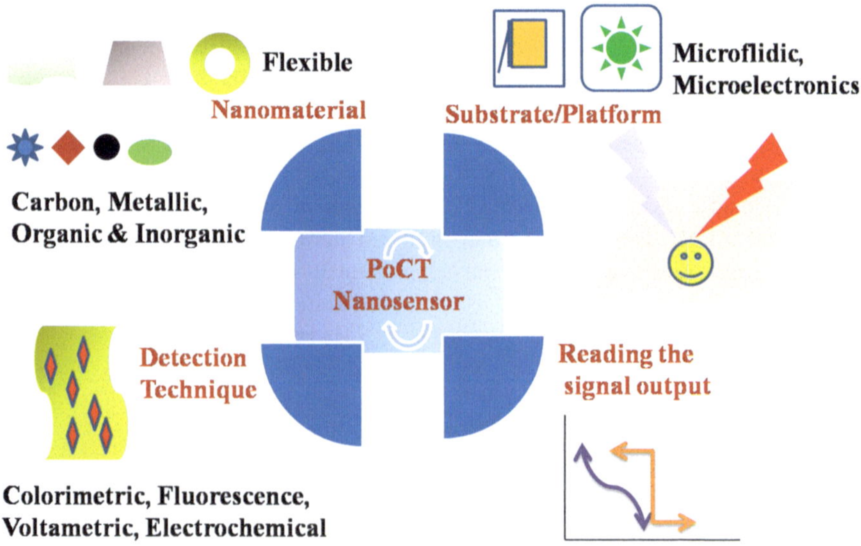

Fig. 4.1 Schematic representation of various components and techniques of PoCT nanosensors

the biological component and physicochemical detecting tool to confirm the presence of analyte in a given sample. Various molecular imaging tools have been used for diagnosing the diseases, but unfortunately, most of them have some limitations and need improvements (Walia et al. 2017). In this, using bionanotechnology has led to a major development in the succession of nanobiosensor and has been successfully applied for biomedical diagnosis. This chapter summarizes the advancements in disease diagnostic tools, mainly through detecting molecular biomarkers, like nucleic acids and proteins, by the use of nanobiosensors.

4.2 Disease-Based Sensing of Target Analytes/Biomarkers

Nanobiosensors are being extensively used for molecular recognition, detection, and sensing of biomarkers which can be used for the diagnosis of particular disease (Chamorro-Garcia and Merkoçi 2016). There are different modalities for medical diagnosis, viz., bioassay, biosensors, and imaging. Currently available methods of diagnosis have certain limitations like low sensitivity, specificity, and reproducibility. Diagnostic methods must possess good sensitivity and should aid in early detection of diseases to provide better treatment options. In the last few years, advancements made in medicine are remarkable. However, the lack of on-time disease detection techniques or proper monitoring of the same is still the major cause for patient death. Respiratory infections, ischemic heart diseases, bacterial infections like diarrheal disease or tuberculosis, diabetes, etc. are few of the major diseases responsible for death worldwide. But at the same time if these can be detected on time, they can also be prevented. Currently, we have sufficient tools and technologies available for their on-time detection; unfortunately also a lot of people are still dying because of these diseases. There can be few probable reasons, viz., short of equipment or required apparatus, high cost, the time-lapse between symptoms noticeable to the patient, unavailability of trained manpower/specialists to run the equipments, etc. (Vidic et al. 2019; Quesada-González and Merkoçi 2018; Saylan et al. 2019).

In this chapter we have thoroughly elaborated the potential of various NMs which can be used as PoC diagnostic tools and discussed their mechanism on how these can be used as innovative materials that can help to miniaturize, support, and improve the excellence of diagnostic devices and technologies. Table 4.1 shows different types of research area based on recent advancements in NMs for PoCT applications.

4.2.1 Nanobiosensors for Diabetes

Type II diabetes, which is a metabolic disorder of glucose dysregulation, also known as hyperglycemia, is the most common form of diabetes (Gu et al. 2013). In one of the study, an innovative nanocomposite system based on ferrocenyl iminopropyl-modified silica nanoparticle conjugates (fap-SiNPs), entrapped in glucose oxidase (GOx) and bovine serum albumin cross-linked with glutaraldehyde, was developed. The system was found to be efficient for electrochemical glucose monitoring with a detection limit of 0.68 nM, and it was able to detect in a linear range from 1 nM to 100 μM. The real-time sample analysis was done using saliva samples (Saadaoui et al. 2016). In another study, a simple method has been developed for synthesizing graphene quantum dots (GQD)-capped silver nanoparticles (AgNPs). The developed nanocomposite consisting of AgNPs/GQDs was characterized by different spectroscopic and microscopic techniques. Authors have proposed this system as a colorimetric sensor which acts as signal probe for detecting hydrogen peroxide (H_2O_2) and glucose. The developed sensing system provides excellent sensitivity and specificity with limit of detection (LOD) of 162 nM and 30 μM for H_2O_2 and glucose sensing, respectively (Nguyen et al. 2018). In another report, researchers have designed a fluorescent carbon dot (CD) surface by benzenediboronic acid (BDBA), which acts as functional ligand and forms boronate complexes with the nanoparticles. This nanosystem was used for the detection of hydrogen peroxide and glucose. The benzenediboronic acid undergoes oxidation in the presence of hydrogen peroxide and correspondingly quenches the fluorescence of CDs through electron transfer process. This simple mix-and-detect strategy was utilized for glucose sensing specifically with detection limit of 0.4 μM. Furthermore, the developed sensing system did not show any change in fluorescence with any other interfering species and can be used in determining glucose in human serum too (Pan et al. 2019). Similarly, researchers have synthesized a simple, economical, and novel glucose oxidase (GOx)-mediated scaffold for sensing glucose based on CDs and AgNPs as nanocomplexes. This system followed three consecutive events: fluorescence quenching of CDs by AgNPs due to surface plasmon-enhanced energy transfer (SPEET) from CDs (as donor) to AgNPs (as acceptor), simultaneous production of

Table 4.1 Different types of advanced NMs and their applications in PoCT-based biomarker sensing

S. no.	NMs	Technique	Disease	Biomarker/Analyte	LOD	Reference
1.	Azide- and alkyne-AuNP	Colorimetric	*E. coli* infection	*E. coli*	40 cfu/mL	Mou et al. (2019)
2.	Surface-functionalized AuNPs	Olfactory (supramolecular-based biosensor)	*E. coli* infection	Lipase	10^2 cfu/mL	Duncan et al. (2017)
3.	Peptide (P1)-modified AuNPs	Quartz crystal microbalance (QCM) biosensor	Type IV collagenase-relevant disease	Type IV collagenase	0.96 $ngmL^{-1}$	Dong et al. (2018)
4.	Multifunctional hybrid graphene oxide-based NM	SERS	Alzheimer's disease	β amyloid and tau protein	100 fg/mL	Demeritte et al. (2015)
5.	PDA (polydopamine)-Apt-GAMMS (glucose-aminated magnetic mesoporous silica) nanocomposites	Phosphoglucomutase (PGM)-based activity assay	Food safety monitoring	Aflatoxin B1 (AFB1)	0.02 ng mL^{-1}	Wang et al. (2019)
6.	Cellulose SERS strips (CS) decorated with graphene-isolated-Au-nanocrystals (GIANs)	SERS	Newborn jaundice	Blood Bilirubin	7 μM	Zou et al. (2018)
7.	Antibody-conjugated magnetic nanobeads (MagNB-Abs) and gold nanozymes (AuNZ-Abs) (MagNBs)	MagLISA	Acute infectious diseases	Influenza virus	5.0×10^{-12} $g{\cdot}mL^{-1}$ by human eyes 44.2×10^{-15} $g{\cdot}mL^{-1}$ by a microplate reader, 2.6 $PFU{\cdot}mL^{-1}$ in clinical samples	Oh et al. (2018)

(continued)

Table 4.1 (continued)

S. no.	NMs	Technique	Disease	Biomarker/Analyte	LOD	Reference
8.	Photoluminescent carbon dots (CDs) embedded in bacterial cellulose (BC) nanopaper (CDBN) Smartphone-based BR assay kit	Visual monitoring	Jaundice	Bilirubin	0.19 mg dL^{-1}	Tabatabaee et al. (2019)
9.	Au-Ag alloy NBs as plasmonic nanostructures and nanoyeast-scFvs as affinity reagents	SERS	Cancer biomarker	Soluble programmed death 1 (sPD-1), soluble programmed death-ligand 1 (sPD-L1), soluble epithermal growth factor receptor (sEGFR)	6.17 pg/mL, 0.68 pg/mL, and 69.86 pg/mL, respectively	Li et al. (2018)
10.	4-Mercaptobenzoic acid-encoded AuNPs	SERS	Blood cancer	Circulating tumor cells (CTCs) in the blood	5 cells/mL	Wu et al. (2015)
11.	Bovine serum albumin-gold nanoclusters (BSA-AuNCs)	Fluorescent assay based on inner filter effect (IFE)	Hypercholesterolemia	Sensing H_2O_2 and cholesterol	0.8 μM and 1.4 μM, respectively	Chang and Ho (2015)
12.	Pyrimidine-based fluorescent probe (R)-4-(anthracen-9-yl)-6-(naphthalen-1-yl)-1,6-dihydropyrimidine-2-amine (ANDPA) tagged with glucose-AgNPs	Fluorescent immunoassay	*Pseudomonas aeruginosa*	*Pseudomonas aeruginosa*	1.5 cfu/mL	Ellairaja et al. (2017)
13.	PEG-gold nanorod	LSPR	Alzheimer's disease	Tau (τ) protein	1.56–1.66 pM	Kim et al. (2019b)

14.	Raman dye-coded polyA aptamer-AuNP (PAapt-AuNP) conjugates	SERS	Alzheimer's disease	Aβ (1–42) oligomers and tau protein	3.7 × 10–2 nM and 2 × 10–4 pM, respectively	Zhang et al. (2019)
15.	Formate-bridged AgNPs and Tb(III) nanosensor	Fluorescence	Neurodegenerative disease	Dopamine (DA)	0.15 nM	Li et al. (2017)
16.	Fluorescent gold nanoclusters (BSA-Au NCs)	Fluorescence quenching	Neurodegenerative disease	L-dopamine	0.622 nM	Govindaraju et al. (2017)

H_2O_2 from oxidation of glucose (GOx as catalyst), and H_2O_2-induced etching of AgNPs. The H_2O_2 produced helps in converting the AgNPs to silver ions. Finally, CDs detached from the surface of AgNPs and re-establish the inherent fluorescence emission profile of CDs. Thus, the enhancement in fluorescence profile directly depends on the amount of H_2O_2 produced, which ultimately depends on the concentration of glucose present. This methodology was efficient for quantitative analysis of glucose, providing detection limit of 1.39 μM. Moreover, this method was also found useful for the accurate assay of glucose in human serum samples (Ma et al. 2017). Recently, a noninvasive hyaluronate-gold nanoparticles (AuNPs)/GOx complex was used for continuous monitoring and measuring of glucose levels in the body fluids. This wireless sensing system showed quick response within 5 s and offered high sensitivity and selectivity with low detection limit of 0.5 mg/dL (Kim et al. 2019a).

4.2.2 Nanobiosensor for Cardiovascular Diseases

Cardiovascular diseases (CVDs) and stroke are the topmost diseases which are responsible for sudden causes of death globally (Mc Namara et al. 2019). According to analysis by the World Health Organization (WHO), in 2015, almost 17.7 million deaths, which are approximately 31% of allover global deaths, were due to heart-related disorders only (Vashistha et al. 2018). Due to this very high prevalence of CVDs, prevention and management of CVDs increasingly demand effective diagnostic testing (King Kevin et al. 2016). Electrocardiography (ECG) with integrated nanostructure conformal antenna acting as wireless transmitter is being widely used for recording spontaneous electrical activity of the heart (Satvekar et al. 2014). CVD progression can only be prevented by regular monitoring of various risk factors, viz., level of total cholesterol (TC), triglyceride (TG), high-density lipoprotein (HDL), and low-density lipoprotein (LDL) (Tada et al. 2018). However, recording these risk factors is generally expensive and depends on high-end equipments. To combat this, several rapid, easy, and simple NM-based biosensors are being designed for regularly testing the level of risk-associated factors. Innovative NMs offer higher stability, biocompatibility, and exceptional physical, chemical, and electronic properties, which hold huge capabilities for expanding PoC diagnostic techniques for CVD. Wu et al. reported the sensing of TG level by developing AuNPs/poly (3,4-ethylenedioxythiophene)-poly(styrene-sulfonate) (PEDOT:PSS) composite-based amperometric biosensor. This biosensing system showed enhanced catalytic activity and electrochemical conductivity as well as offered improved sensitivity with a very fast response time (within seconds) and LOD of 7.88 $mgdL^{-1}$ (Wu et al. 2014). In another report, TG biosensor was fabricated by lipase and nanoporous gold (NPG) composite. The developed sensing tool was able to specifically detect TG ranging from 50 $mgdL^{-1}$ to 250 $mgdL^{-1}$. Authors have also suggested that this biosensor could be employed for monitoring TG in human serum (Lu et al. 2017). Furthermore, increased level of TC is well-known to cause CVD; hence measuring its level by using biosensing tool is important for the management and treatment of

CVD. Many researchers have documented its sensing activity principle which is based on measuring the level of H_2O_2 produced in TC catalysis by using cholesterol oxidase (ChOx) as well as cholesterol esterase (ChE). Various types of NMs, viz., nanostructured platinum, NPG networks, and AuNPs, were applied to fabricate and develop electrochemical biosensor for measuring TC with enhanced analytical performance (Lu et al. 2017). Quantum dots (QDs), because of their inherent properties, viz., energy transfer process between QDs and acceptor moiety, high fluorescence, and photostability, are being widely used in monitoring the LDL level. Researchers have prepared L-cysteine-capped CdS QDs deposited on Au substrate and conjugated it with apolipoprotein B-100 (AAB) antibodies. It was found that CysCdS QDs increased the sensitivity of the developed composite for LDL sensing in a linear range from 5 mg dL^{-1} to 120 mg dL^{-1} by both electrochemical and surface plasmon resonance (SPR) techniques (Ali et al. 2014). Authors have proposed a selective, reproducible, and biofunctionalized QD-nickel oxide (NiO) nanorod-based platform for lipid detection. The label-free AAB-conjugated CdS QD and NiO nanocomposite system act as a platform for the detection of lipid profile in human serum samples. The biosensing system showed selectivity and high sensitivity toward LDL with a low detection limit of 0.05 mg dl^{-1} (Ali et al. 2016). It has been reported that high-density lipoprotein (HDL) detection can be done by two approaches, viz., by measuring HDL cholesterol as well as by monitoring the HDL particles. For this, combination of an electrochemical sensor and a silver paste electrode was used for measuring HDL cholesterol via homogeneous assay-based biosensor. This system was used to quantify HDL cholesterol using a polyoxyethylene alkylene tribenzylphenyl ether surfactant (Emulgen B-66). They have confirmed that Emulgen B-66 enhanced the electrocatalytic reduction of H_2O_2 as well as dissolved the HDL particles selectively. This sensing tool showed a linear response in concentration range varying between 0.5 and 4 mM (Ahmadraji et al. 2014).

4.2.3 Nanobiosensor for Neurodegenerative Diseases

Currently more than 10% of people worldwide of age 65 years are suffering from Alzheimer's disease (AD), a neurodegenerative disease related to aging, resulting in dementia. This number may reach to 100 million by the end of 2050 (Kwon et al. 2016; Robert et al. 2015). Because of very less available diagnostic techniques, medical practitioners lack experience in analyzing clinical diagnosis of AD (Sabbagh et al. 2017). Biocompatible NMs having the ability to cross the blood-brain barrier can be utilized for improving the diagnosis of various neurodegenerative diseases in their premature stages by means of an easy, economical, and precise strategy (Leszek et al. 2017). NMs, such as electrochemically reduced GOx and gold nanowires (GNWs), have reported very good sensitivity of 1.7 fM for miR-137, which is known to be a confirmed and established biomarker for the early diagnosis of AD. The applicability of this nanobiosensing tool was also tested in a real sample of human plasma (Azimzadeh et al. 2017; Ma et al. 2017). In another literature report, authors have developed an exfoliated graphene oxide (EGO)- and GNW-based electrochemical nanobiosensor for early detection of Parkinson's disease (PD) which detects the circulating biomarker, miR-195. This biosensing tool was able to detect miR-195 with very low detection limit of ~2.9 fM. The nanosystem was found reliable even with real samples having high sensitivity and detection under dynamic range of 10–900 fM. Additionally, this system was found selective for specific detection of miRNA over nonspecific oligonucleotides. Hence authors have suggested that the miR-195 electrochemical nanobiosensor could be a good choice for clinicians for the medical diagnosis of PD (Aghili et al. 2018). Vilela et al. have prepared GOx/upconversion NP sensors which were utilized for the detection of oligonucleotide sequences relevant to mRNA associated with AD and prostate cancer. The sensing system was found to be highly effective, sensitive, as well as selective for mRNA biomarkers which can sense throughout the linear range varying from 200 fM to 5 nM. The developed system was photostable and did not show any biomolecule auto-absorbance interferences (Vilela et al. 2016). Haes et al. demonstrated a localized surface plasmon resonance (LSPR) spectroscopy-based optical nanobiosensor. This sensing system was made up of triangular-shaped AgNPs deposited on mica substrates to examine the interaction between the antigen, amyloid β-derived diffusible ligands (ADDLs), and specific anti-ADDL antibodies. This sandwich assay-based nanosensor was useful to provide quantitative binding information for both antigen and detection of secondary antibody which further allows the monitoring of ADDL concentration. This methodology offered the quantification of the aggregation mechanisms of the AD at physiologically relevant monomer concentrations. Authors have quantified the binding constants for both concentrations, viz., below 10 pM and between 10 and 100 pM, to be $7.3 \times 10^{12}\ M^{-1}$ and $9.5 \times 10^{8}\ M^{-1}$, respectively. Thus, the system was found to be very selective and sensitive toward the detection of ADDL with LOD of <100 fM (Haes et al. 2005). Similarly, researchers have explored the biosensing potential of fluorescent dye Alexa 647 and cadmium selenide/zinc sulfide (CdSe@ZnS) QDs for the detection of apolipoprotein E (ApoE) for targeting AD. The QDs' analytical performance was found to vary at different excitation wavelengths. The QDs at wavelength of 633 nm provide LOD of $\sim$247 pg mL^{-1}, whereas, when the excitation wavelength changed to 532 nm, sensitivity significantly improved as it reaches to $\sim$62 pg mL^{-1}. The developed microarray was found to be 5–7 times more sensitive than that of the Alexa microarray and ELISA. Authors suggested that this strategy could be applied for the detection of different ApoE isoforms and additional AD-related biomarkers (Morales-Narváez et al. 2012).

4.2.4 Nanobiosensor for Jaundice

Heme oxygenase is the primary enzyme which is responsible for catabolism of heme into bilirubin by two major events, first by converting heme to biliverdin and subsequently converting biliverdin to bilirubin in the presence of biliverdin reductase. Bilirubin after binding with albumin proteins circulates in the blood and is then transported to the liver. Higher concentration of bilirubin in its free form causes toxicity and hence acts as a biomarker for the diagnosis of jaundice which further allows us to assess proper liver function and chronic kidney disease in patients (Daood et al. 2009; Tsai and Tarng 2019). The normal human serum concentration level of free bilirubin is <25 μmol/L, whereas in jaundice condition the level may increase to >50 μmol/L. Under impaired hepatic function, free bilirubin cannot be eliminated out of the body, which may result in jaundice and further can lead to brain damage or even death. Thus, it is of utmost importance to monitor bilirubin concentration in human serum for diagnosing jaundice and liver disorders (Du et al. 2017; Feng et al. 2013; Wei et al. 2012). In literature, bilirubin detection was performed by a PoC electrochemical sensor, synthesized by functionalization of NMs on screen-printed carbon electrodes (SPEs). In this electrochemical sensor, the main core consists of multiwalled carbon nanotubes (MWCNT) and graphene. Biliverdin, the oxidized form of bilirubin, was analyzed at potential of +0.25 V along with a peak at +0.48 V, depicting the electrochemical conversion of biliverdin into purpurin. The results revealed that approximately threefold higher sensitivity was observed for graphene NMs (0.1 ± 0.018 nM) as compared to MWCNT (0.3 ± 0.022 nM) (Thangamuthu et al. 2018). It is already known that many types of hepatic diseases and blood disorders can be diagnosed by determining the serum bilirubin concentration. To this, fluorescent sensing of free bilirubin was monitored by a polydopamine/gold nanocluster composite film. It was found that, when the composite film was monitored in the presence of free bilirubin, fluorescence quenching was observed. This resulted due to the strong complexation between bilirubin and BSA. Thus, the system was found to be fast, eco-friendly, and highly selective, having LOD for free bilirubin ~0.61 μmol/L (Li et al. 2019). In another report, bilirubin biosensor was prepared by immobilizing bilirubin oxidase (BOx) on graphene oxide NPs (GONPs). These GONPs were earlier fabricated with polypyrrole (Ppy) which was electrochemically deposited onto fluorine-doped tin oxide (FTO) plate. This amperometric biosensor was found to be very sensitive and selective toward bilirubin detection, when polarized at +0.2 V under ambient conditions. The LOD and limit of quantification (LOQ) for bilirubin sensing was found to be 0.1 nM and 2.6 nM, respectively. The developed nanohybrid system was highly stable over a period of 5 months and was reusable when stored at 4 °C (Chauhan et al. 2016). In another report, graphene-isolated-Au-nanocrystal-loaded cellulose paper strips (CS@GIANs) were fabricated to detect bilirubin in blood samples. This SERS-based sensitive nanoplatform showed selective binding with bilirubin and offered LOD of 7 μM (Zou et al. 2018).

4.2.5 Nanobiosensor for Microbial Infection

The world is continuously facing growing concerns over the microbial infections, which include both bacterial and fungal infection. The resistant strains are rapidly increasing the designing of innovative therapeutic molecules day by day. Each year almost 300 million cases of bacterial infections are being observed. The fungal infections can be harmful for both humans and agriculture. Therefore, there is an urgent need for development of modern diagnostic and detection tools for microbial infections, and this area is becoming one of the most important in nanomedicine-based healthcare system (Hutnick and Pokorski 2018; Miranda et al. 2011). Duan et al. reported simultaneous detection of two pathogenic bacteria, viz., *Vibrio parahaemolyticus* and *Salmonella typhimurium*. This innovative biosensing technique was based on the principle of fluorescence resonance energy transfer (FRET). Here, the green-emitting QDs (gQDs) and red-emitting QDs (rQDs) acted as donors, whereas carbon nanoparticles (CNPs) were the electron acceptor moiety. The fluorescence intensity exhibited by QDs was found to vary linearly with the concentration of the pathogens, ranging from 50 cfu mL^{-1} to 106 cfu mL^{-1}, with detection limits of 25 and 35 cfu mL^{-1} for *V. parahaemolyticus* and *S. typhimurium*, respectively (Duan et al. 2015). Furthermore, thionine-labeled anti-*Aeromonas hydrophila* (anti-Ah) and ferrocene-labeled anti-*Pseudomonas aeruginosa* (anti-Ps) redox probes were attached with the electrochemical immunosensor for detecting these two bacteria, respectively. In this system, zeolitic imidazolate framework/AuNP composite was used as a platform for detection of the bacteria. This nanocomposite platform was able to detect *P. aeruginosa* and *A. hydrophila* linearly in the range of $10–10^5$ and $10–10^7$ cfu mL^{-1}, with LOD of 3.53 and 3.61 cfu mL^{-1}, respectively (Balaji Viswanath et al. 2018). The food-borne pathogens have been identified as the global cause of the majority of infections. Fu et al. developed a simple and fast method, based on polymerase chain reaction (PCR), where colorimetric AuNPs were used for the detection of *Listeria monocytogenes* and *Salmonella enterica*. The sulfur-gold linkage in AuNP-PCR complex products was found to be more stable than primers linked to AuNPs. This method was found to be capable of analyzing genomic DNA of *L. monocytogenes* and *S. enterica* specifically with LOD of 0.015 ng mL^{-1} and 0.013 ng mL^{-1}, respectively (Fu et al. 2013). In another report, researchers have developed a colorimetric conjugated system of enzyme and AuNPs for the detection of microbial contamination by choosing *Escherichia coli (XL1)* as a model analyte microorganism. The NM composed of cationic AuNP-bound β-galactosidase (β-Gal) resulted in the inhibition of the activity of β-galactosidase. *E. coli*, having affinity for binding with these NPs, simultaneously releases the β-Gal, and the enzyme regains its inherent activity. By using this, 10^2 bacteria/mL in solution form and 10^4 bacteria/mL in a test strip were detected (Miranda et al. 2011).

4.2.6 Nanobiosensor for Cancer

Cancer is one of the severe diseases, which is the second leading cause of death (Siegel et al. 2019). Inorganic and organic NMs have been used as potential bioimaging and targeted drug delivery systems against various cancers. A large number of NMs are in the process of clinical trials for the treatment of different types of cancers (Zottel et al. 2019; Bobo et al. 2016; Tran et al. 2017; Jabir et al. 2018). Ultrasensitive and precise PoC cancer diagnostic tools are required for the early detection and screening of cancer biomarkers with more accuracy and specificity (Chen and Zhao 2019; Huddy et al. 2015). Cancer biomarkers may include methylated DNA, proteins like prostate-specific antigen (PSA), mutations in tumor suppressor genes (p53), micro-RNA (miRNA), etc. (Van der Drift et al. 2010; Pagel et al. 2015; Gaikwad and Banerjee 2018). Zheng and coworkers designed self-targeting fluorescent CDs for diagnosing brain cancer cells. The researchers synthesized novel type of biocompatible, full color-emitting CDs made up of D-glucose and L-aspartic acid which can target brain cancer glioma. In vivo imaging results of orthotopic C6 glioma-bearing mice indicated that these developed NMs can be used as high-contrast imaging agent specifically in brain site (Zheng et al. 2015). In another report, a green synthesis approach was used to synthesize folic acid (FA)-conjugated $MnCo_2O_4$ nanotubes. These nanotubes showed bimodal diagnosis capabilities, i.e., pressure- and colorimetric-based detection of HeLa cancer cells. Authors claimed that this system mimic the artificial peroxidase enzyme, which generates O_2 as pressure signal, and 3,3′,5,5′-tetramethylbenzidine (TMB) will be produced as a blue-colored complex. Figure 4.2 depicts the specific detection of cancer cells on the basis of changes in the absorbance and pressure signals. On the basis of linear relationship between pressure, absorbance, and cell number, the LOD was found to be 50 cells/mL (Ding et al. 2018). Carcinoembryonic antigen (CEA) is one of the important tumor biomarkers for early-stage cancer diagnosis (Ranga et al. 2016; Demirci Uzun et al. 2014; Viter et al. 2016). In one study, a composite of AuNPs and nickel hexacyanoferrate NPs (NiHCFNPs) deposited on glassy carbon electrode (GCE) was prepared. This composite was used for the immunological determination of CEA in clinical samples. Under optimal conditions, the peak current of the immunosensor decreased linearly with increasing CEA concentrations. The LOD was found to be 0.1 ng/mL (Yuan et al. 2009). Graphene is one of the excellent NM for sensing the cancer biomarkers because it provides outstanding conductivity as well as catalytic activity (Chauhan et al. 2016; Pandikumar et al. 2014; Pumera et al. 2010; Chu et al. 2015). Graphene platform was utilized for electrochemical biosensing of CEA. For this, the graphene was conjugated with the magnetic beads (MBs) and enzyme-labeled antibody-AuNP bioconjugate. In this system, the nanostructures of MBs provided high surface area, fast kinetics, and stability because of their small size. The developed sensing system showed LOD of 5 ng mL^{-1}. Since it is easy to modify MBs with suitable functional groups, these have been widely used in different types of electrochemical immunoassay systems (Xie and Jon 2012; Jin et al. 2014). In another study, MBs were initially coated with primary antibodies (Ab_1) and subsequently combined with graphene sheets by providing magnetic field externally. After that, target antibody (Ab_2) was combined with HRP-fabricated AuNPs (Ab_2-AuNPs-HRP). By using these highly sensitive NMs, the electron transfer between electrode and PSA was accelerated. The sensitivity and LOD value for PSA detection were found to be 31.5 μA mL ng^{-1} and 0.5 pg/mL, respectively (Mani et al. 2009, 2011). Graphene composite of Au and Ag was developed as an immunoassay sensor for the detection of CEA. In this system, AuNPs and AgNPs offered electrocatalytic activity which ultimately amplified the signal for the immunosensor. The detection limit for CEA was found to be 8 pg/mL,

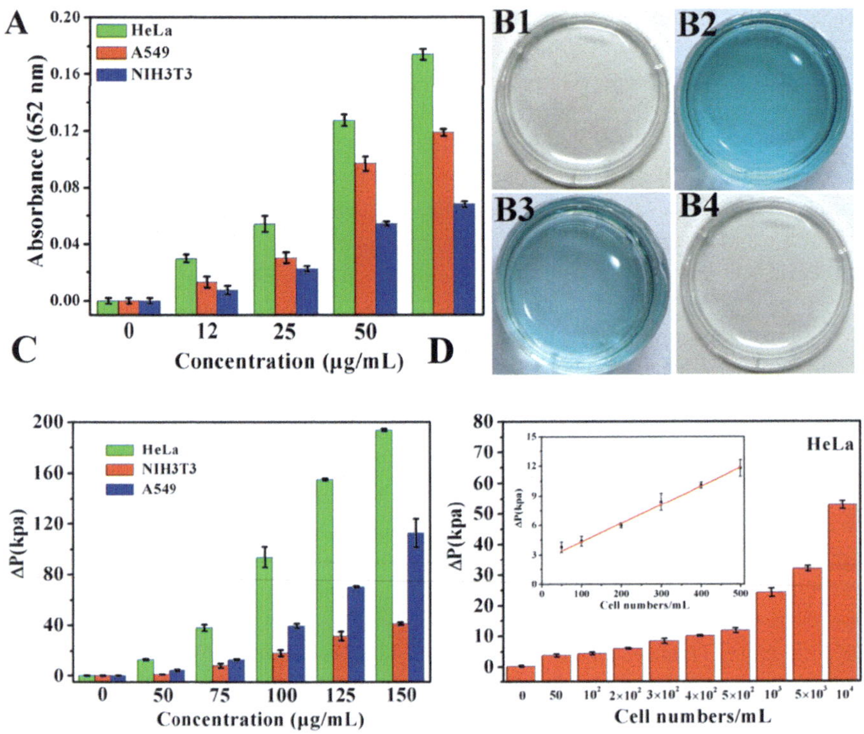

Fig. 4.2 (**a**) Absorbance profile of cells at 652 nm after incubation with FA-$MnCo_2O_4$ nanotubes for 1 h, followed by addition of 16 mM TMB in buffer of pH = 5.0. Images of HeLa cells incubated with TMB and buffer in absence of nanotubes (**b1**), HeLa cells (**b2**), A549 cells (**b3**), and NIH3T3 cells (**b4**) incubated with 100 μg mL^{-1} FA-$MnCo_2O_4$ nanotubes for 1 h and then addition of TMB in 200 μL of buffer. (**c**) Pressure-change profiles of cells after incubation with FA-$MnCo_2O_4$ nanotubes for 1 h and then addition of H_2O_2. (**d**) Quantitative detection of HeLa cells using FA-$MnCo_2O_4$ nanotubes and H_2O_2 on the basis of pressure change. (Adapted from Ding et al. (2018), reprint with permission)

and it can be detected linearly in the range between 10 and 1.2×10^5 pg/mL (Huang et al. 2015). A bifunctional nanocomposite probe was developed, which demonstrated detection of CEA via magnetic focusing and surface-enhanced Raman scattering (SERS) in a microfluidic platform. This nanocomposite probe mainly consisted of magnetic nickel-iron core and a gold shell. This system offered detection sensitivity, with LOD of ~0.1 pM (Li et al. 2015). Cytosensors are the different types of biosensors which sense live cells as the recognizing elements. Beside detecting and sensing cancer biomarkers, direct electrochemical detection of cancerous cells is also an alternative approach to use the electrochemical technique-based biosensing in these areas (Seven et al. 2013; Laocharoensuk 2016). Another study was carried out where folate receptor-rich HeLa cells were detected with Au electrode which was previously modified with FA functionalized AuNPs. For this cytosensing tool, after immobilizing AuNPs onto Au electrode,

11-mercaptoundecanoic acid (MUA) was further conjugated. The FA which mainly recognizes HeLa cells was then combined with MUA coating, and finally, Au electrode/AuNP/MUA/FA complex system was formed. This nanosystem was used for detection of FA-rich HeLa cells via electrochemical impedance spectroscopy (EIS). Electrode impedance was increased after the adsorption of HeLa cells on the electrode. The nanosystem was found to be highly selective and sensitive toward HeLa cells, with LOD of 6 cells/mL (Wang et al. 2012; Anik and Timur 2016).

4.3 Concluding Remarks and Future Perspectives

Although a lot of efforts for the treatment of various diseases have been made to avoid epidemics and high mortality rate, unfortunately diagnosing the diseases at the early stages is still one of the major healthcare concerns. Thus, developing more sensitive, specific, fast, accurate, economical, and user-friendly diagnostic tools is indeed an urgent demand. However, like any other emerging field, nanobiosensing also face many challenges, viz., amplifying the signals and enhancing the signal to noise ratio and their transduction. Biomolecules having unique structures possess specific functions. Hence assessing their use in detection and designing specific NMs is still a great challenge. In fact, the mechanism of interaction between biomolecules and NMs is also not yet fully understood. In this chapter, we have provided an overview of currently developed NMs and their use in biosensing applications for different diseases like jaundice, cardiovascular, neurodegenerative, diabetes, cancer, and microbial infections, with an emphasis on the nanotechnology-based PoCT system. Also, details have been given on various novel NMs and their corresponding nanocomposites for sensing respective biomarkers for different diseases. Though NM-based biosensing systems have numerous advantages like high specificity, sensitivity, easy availability, portability, and disposability over conventional systems, they all are still in the early phase of their development. At this time, the most promising PoCT devices may perform well in some ways, but they might not be successful to convince all the criteria. Therefore, for the development of highly integrated and novel NM-based PoCT system, substantial efforts are required from scientific fields of biology, chemistry, physics, engineering, and medical perspectives, which can ultimately provide the support for management and prevention of various dreadful diseases. In the future, attention should be given for fabricating a new generation of biosensors and also clarifying the mechanism of interactions between the NMs and biomolecules.

Acknowledgment The authors would like to thank the director of CSIR-IHBT for his constant support and encouragement. AA acknowledges financial assistance in the form of project grant MLP-0201 from CSIR and GAP-0214 (EMR/2016/003027) from DST, government of India. CS and SW acknowledge CSIR for their respective SRF fellowship. The CSIR-IHBT communication number for this manuscript is 4530.

References

Aghili Z, Nasirizadeh N, Divsalar A, Shoeibi S, Yaghmaei P (2018) A highly sensitive miR-195 nanobiosensor for early detection of Parkinson's disease. Artif Cells Nanomed Biotechnol 46 (1):32–40

Ahmadraji T, Gonzalez-Macia L, Killard AJ (2014) A biosensor for the determination of high density lipoprotein cholesterol employing combined surfactant-derived selectivity and sensitivity enhancements. Anal Methods 6(12):3975–3981

Ali MA, Srivastava S, Pandey MK, Agrawal VV, John R, Malhotra BD (2014) Protein–conjugated quantum dots interface: binding kinetics and label-free lipid detection. Anal Chem 86 (3):1710–1718

Ali MA, Srivastava S, Agrawal VV, Willander M, John R, Malhotra BD (2016) A biofunctionalized quantum dot–nickel oxide nanorod based smart platform for lipid detection. J Mater Chem B 4 (15):2706–2714

Anik U, Timur S (2016) Towards the electrochemical diagnosis of cancer: nanomaterial-based immunosensors and cytosensors. RSC Adv 6(113):111831–111841

Azimzadeh M, Nasirizadeh N, Rahaie M, Naderi-Manesh H (2017) Early detection of Alzheimer's disease using a biosensor based on electrochemically-reduced graphene oxide and gold nanowires for the quantification of serum microRNA-137. RSC Adv 7(88):55709–55719

Bajpai VK, Kamle M, Shukla S, Mahato DK, Chandra P, Hwang SK, Han YK (2018) Prospects of using nanotechnology for food preservation, safety, and security. J Food Drug Anal 26 (4):1201–1214

Balaji Viswanath K, Krithiga N, Jayachitra A, Sheik Mideen AK, Amali AJ, Vasantha VS (2018) Enzyme-free multiplex detection of Pseudomonas aeruginosa and Aeromonas hydrophila with ferrocene and thionine-labeled antibodies using ZIF-8/Au NPs as a platform. ACS Omega 3 (12):17010–17022

Balakrishnan SR, Hashim U, Gopinath SC, Poopalan P, Ramayya HR, Omar MI, Veeradasan P (2015) A point-of-care immunosensor for human chorionic gonadotropin in clinical urine samples using a cuneated polysilicon nanogap lab-on-chip. PLoS One 10(9):e0137891

Bobo D, Robinson KJ, Islam J, Thurecht KJ, Corrie SR (2016) Nanoparticle-based medicines: a review of FDA-approved materials and clinical trials to date. Pharm Res 33(10):2373–2387

Cammann K (1977) Bio-sensors based on ion-selective electrodes. Fresenius' Z Anal Chem 287 (1):1–9

Cassano D, Pocovi-Martinez S, Voliani V (2017) Ultrasmall-in-nano approach: enabling the translation of metal nanomaterials to clinics. Bioconjug Chem 29(1):4–16

Chamorro-Garcia A, Merkoçi A (2016) Nanobiosensors in diagnostics. Nano 3:1849543516663574

Chang HC, Ho JAA (2015) Gold nanocluster-assisted fluorescent detection for hydrogen peroxide and cholesterol based on the inner filter effect of gold nanoparticles. Anal Chem 87 (20):10362–10367

Chauhan N, Rawal R, Hooda V, Jain U (2016) Electrochemical biosensor with graphene oxide nanoparticles and polypyrrole interface for the detection of bilirubin. RSC Adv 6 (68):63624–63633

Chen M, Zhao H (2019) Next-generation sequencing in liquid biopsy: cancer screening and early detection. Hum Genomics 13(1):34

Christodouleas DC, Kaur B, Chorti P (2018) From point-of-care testing to eHealth diagnostic devices (eDiagnostics). ACS Cent Sci 4(12):1600–1616

Chu Z, Liu Y, Xu Y, Shi L, Peng J, Jin W (2015) In-situ fabrication of well-distributed gold nanocubes on thiol graphene as a third-generation biosensor for ultrasensitive glucose detection. Electrochim Acta 176:162–171

Chung IM, Rajakumar G, Gomathi T, Park SK, Kim SH, Thiruvengadam M (2017) Nanotechnology for human food: advances and perspective. Front Life Sci 10(1):63–72

Cui Y (2017) Wireless biological electronic sensors. Sensors 17(10):2289

Daood MJ, McDonagh AF, Watchko JF (2009) Calculated free bilirubin levels and neurotoxicity. J Perinatol 29(S1):S14

Demeritte T, Viraka Nellore BP, Kanchanapally R, Sinha SS, Pramanik A, Chavva SR, Ray PC (2015) Hybrid graphene oxide based plasmonic-magnetic multifunctional nanoplatform for selective separation and label-free identification of Alzheimer's disease biomarkers. ACS Appl Mater Interfaces 7(24):13693–13700

Demirci Uzun S, Kayaci F, Uyar T, Timur S, Toppare L (2014) Bioactive surface design based on functional composite electrospun nanofibers for biomolecule immobilization and biosensor applications. ACS Appl Mater Interfaces 6(7):5235–5243

Ding X, Yang KL (2013) Development of an oligopeptide functionalized surface plasmon resonance biosensor for online detection of glyphosate. Anal Chem 85(12):5727–5733

Ding E, Hai J, Chen F, Wang B (2018) Constructing 2D nanosheet-assembled $MnCo_2O_4$ nanotubes for pressure and colorimetric dual-signal readout detection of cancer cells in serum samples. ACS Appl Nano Mater 1(8):4156–4163

Dong ZM, Jin X, Zhao GC (2018) Amplified QCM biosensor for type IV collagenase based on collagenase-cleavage of gold nanoparticles functionalized peptide. Biosens Bioelectron 106:111–116

Du Y, Li X, Lv X, Jia Q (2017) Highly sensitive and selective sensing of free bilirubin using metal–organic frameworks-based energy transfer process. ACS Appl Mater Interfaces 9 (36):30925–30932

Duan N, Wu S, Dai S, Miao T, Chen J, Wang Z (2015) Simultaneous detection of pathogenic bacteria using an aptamer based biosensor and dual fluorescence resonance energy transfer from quantum dots to carbon nanoparticles. Mikrochim Acta 182(5–6):917–923

Duncan B, Le ND, Alexander C, Gupta A, Yesilbag Tonga G, Yazdani M, Li X (2017) Sensing by smell: nanoparticle–enzyme sensors for rapid and sensitive detection of bacteria with olfactory output. ACS Nano 11(6):5339–5343

Ellairaja S, Krithiga N, Ponmariappan S, Vasantha VS (2017) Novel pyrimidine tagged silver nanoparticle based fluorescent immunoassay for the detection of Pseudomonas aeruginosa. J Agr Food Chem 65(8):1802–1812

Feng Q, Du Y, Zhang C, Zheng Z, Hu F, Wang Z, Wang C (2013) Synthesis of the multi-walled carbon nanotubes-COOH/graphene/gold nanoparticles nanocomposite for simple determination of Bilirubin in human blood serum. Sens Actuators B Chem 185:337–344

Freeman MH, Hall JR, Leopold MC (2013) Monolayer-protected nanoparticle doped xerogels as functional components of amperometric glucose biosensors. Anal Chem 85(8):4057–4065

Fu Z, Zhou X, Xing D (2013) Rapid colorimetric gene-sensing of food pathogenic bacteria using biomodification-free gold nanoparticle. Sens Actuator B Chem 182:633–641

Gaikwad PS, Banerjee R (2018) Advances in point-of-care diagnostic devices in cancers. Analyst 143(6):1326–1348

Govindaraju S, Ankireddy SR, Viswanath B, Kim J, Yun K (2017) Fluorescent gold nanoclusters for selective detection of dopamine in cerebrospinal fluid. Sci Rep 7:40298

Gu Z, Aimetti AA, Wang Q, Dang TT, Zhang Y, Veiseh O, Anderson DG (2013) Injectable nano-network for glucose-mediated insulin delivery. ACS Nano 7(5):4194–4420

Haes AJ, Chang L, Klein WL, Van Duyne RP (2005) Detection of a biomarker for Alzheimer's disease from synthetic and clinical samples using a nanoscale optical biosensor. J Am Chem Soc 127(7):2264–2271

Huang J, Tian J, Zhao Y, Zhao S (2015) Ag/Au nanoparticles coated graphene electrochemical sensor for ultrasensitive analysis of carcinoembryonic antigen in clinical immunoassay. Sens Actuator B Chem 206:570–576

Huddy JR, Ni MZ, Markar SR, Hanna GB (2015) Point-of-care testing in the diagnosis of gastrointestinal cancers: current technology and future directions. World J Gastroenterol 21 (14):4111

Hutnick MA, Pokorski JK (2018) Polymeric interventions for microbial infections: a review. Mol Pharm 15(8):2910–2921

Jabir NR, Anwar K, Firoz CK, Oves M, Kamal MA, Tabrez S (2018) An overview on the current status of cancer nanomedicines. Curr Med 34(5):911–921

Jeevanandam J, Barhoum A, Chan YS, Dufresne A, Danquah MK (2018) Review on nanoparticles and nanostructured materials: history, sources, toxicity and regulations. Beilstein J Nanotechnol 9(1):1050–1074

Jin B, Wang P, Mao H, Hu B, Zhang H, Cheng Z, Jin Q (2014) Multi-nanomaterial electrochemical biosensor based on label-free graphene for detecting cancer biomarkers. Biosens Bioelectron 55:464–469

Kim SK, Jeon C, Lee GH, Koo J, Cho SH, Han S, Hahn SK (2019a) Hyaluronate-gold nanoparticle/glucose oxidase complex for highly sensitive wireless non-invasive glucose sensors. ACS Appl Mater Interfaces 11(40):37347–37356

Kim H, Lee JU, Kim S, Song S, Sim SJ (2019b) A nanoplasmonic biosensor for ultrasensitive detection of Alzheimer's disease biomarker using a chaotropic agent. ACS Sens 4(3):595–602

King Kevin R, Grazette LP, Paltoo DN, McDevitt JT, Sia SK, Barrett PM, Apple FS (2016) Point-of-care technologies for precision cardiovascular care and clinical research: National Heart, Lung, and Blood Institute Working Group. JACC Basic Transl Sci 1(1–2):73–86

Kwon HJ, Cha MY, Kim D, Kim DK, Soh M, Shin K, Mook-Jung I (2016) Mitochondria-targeting ceria nanoparticles as antioxidants for Alzheimer's disease. ACS Nano 10(2):2860–2870

Laocharoensuk R (2016) Development of electrochemical immunosensors towards point-of-care cancer diagnostics: clinically relevant studies. Electroanalysis 28(8):1716–1729

Leszek J, Md Ashraf G, Tse WH, Zhang J, Gasiorowski K, Fidel Avila-Rodriguez M, Aliev G (2017) Nanotechnology for Alzheimer disease. Curr Alzheimer Res 14(11):1182–1189

Li J, Skeete Z, Shan S, Yan S, Kurzatkowska K, Zhao W, Zhong CJ (2015) Surface enhanced Raman scattering detection of cancer biomarkers with bifunctional nanocomposite probes. Anal Chem 87(21):10698–10702

Li H, Shen J, Cui R, Sun C, Zhao Y, Wu X, Tang B (2017) A highly selective and sensitive fluorescent nanosensor for dopamine based on formate bridged Tb (III) complex and silver nanoparticles. Analyst 142(22):4240–4246

Li J, Wang J, Grewal YS, Howard CB, Raftery LJ, Mahler S, Trau M (2018) Multiplexed SERS detection of soluble cancer protein biomarkers with gold–silver alloy nanoboxes and nanoyeast single-chain variable fragments. Anal Chem 90(17):10377–10384

Li Z, Xiao W, Huang R, Shi Y, Fang C, Chen Z (2019) A gold nanoclusters film supported on polydopamine for fluorescent sensing of free bilirubin. Sensors 19(7):1726

Lu S, Yu T, Wang Y, Liang L, Chen Y, Xu F, Wang S (2017) Nanomaterial-based biosensors for measurement of lipids and lipoproteins towards point-of-care of cardiovascular disease. Analyst 142(18):3309–3321

Ma JL, Yin BC, Wu X, Ye BC (2017) Simple and cost-effective glucose detection based on carbon nanodots supported on silver nanoparticles. Anal Chem 89(2):1323–1328

Mani V, Chikkaveeraiah BV, Patel V, Gutkind JS, Rusling JF (2009) Ultrasensitive immunosensor for cancer biomarker proteins using gold nanoparticle film electrodes and multienzyme-particle amplification. ACS Nano 3(3):585–594

Mani V, Chikkaveeraiah BV, Rusling JF (2011) Magnetic particles in ultrasensitive biomarker protein measurements for cancer detection and monitoring. Expert Opin Med Diagn 5 (5):381–391

Mc Namara K, Alzubaidi H, Jackson JK (2019) Cardiovascular disease as a leading cause of death: how are pharmacists getting involved? Integr Pharm Res Pract 8:1

Mehrotra P (2016) Biosensors and their applications–a review. J Oral Biol Craniofac Res 6 (2):153–159

Miranda OR, Li X, Garcia-Gonzalez L, Zhu ZJ, Yan B, Bunz UH, Rotello VM (2011) Colorimetric bacteria sensing using a supramolecular enzyme–nanoparticle biosensor. J Am Chem Soc 133 (25):9650–9653

Morales-Narváez E, Montón H, Fomicheva A, Merkoçi A (2012) Signal enhancement in antibody microarrays using quantum dots nanocrystals: application to potential Alzheimer's disease biomarker screening. Anal Chem 84(15):6821–6827
Mou XZ, Chen XY, Wang JH, Zhang Z, Yang Y, Shou ZX, Qiu L (2019) Bacteria-instructed click chemistry between functionalized gold nanoparticles for point-of-care microbial detection. ACS Appl Mater Interfaces 11(26):23093–23101
Nayak S, Blumenfeld NR, Laksanasopin T, Sia SK (2016) Point-of-care diagnostics: recent developments in a connected age. Anal Chem 89(1):102–123
Nguyen ND, Van Nguyen T, Chu AD, Tran HV, Tran LT, Huynh CD (2018) A label-free colorimetric sensor based on silver nanoparticles directed to hydrogen peroxide and glucose. Arab J Chem 11(7):1134–1143
Oh S, Kim J, Tran VT, Lee DK, Ahmed SR, Hong JC, Lee J (2018) Magnetic nanozyme-linked immunosorbent assay for ultrasensitive influenza A virus detection. ACS Appl Mater Interfaces 10(15):12534–12543
Pagel O, Loroch S, Sickmann A, Zahedi RP (2015) Current strategies and findings in clinically relevant post-translational modification-specific proteomics. Expert Rev Proteomic 12 (3):235–253
Pan M, Xu Z, Jiang Q, Feng J, Sun J, Wang F, Liu X (2019) Interfacial engineering of carbon dots with benzenediboronic acid for fluorescent biosensing. Nanoscale Adv 1(2):765–771
Pandikumar A, How GTS, See TP, Omar FS, Jayabal S, Kamali KZ, Lim HN (2014) Graphene and its nanocomposite material based electrochemical sensor platform for dopamine. RSC Adv 4 (108):63296–63323
Prasad S (2014) Nanobiosensors: the future for diagnosis of disease. Nanobiosens Dis Diagn 3:1–10
Pumera M, Ambrosi A, Bonanni A, Chng ELK, Poh HL (2010) Graphene for electrochemical sensing and biosensing. Trends Anal Chem 29(9):954–965
Quesada-González D, Merkoçi A (2018) Nanomaterial-based devices for point-of-care diagnostic applications. Chem Soc Rev 47(13):4697–4709
Ranga GS, Muktesh G, Sharma SB, Madhu SV (2016) A study of carcinoembryonic antigen concentrations in patients with coronary artery disease. J Acute Med 6(1):11–18
Rizwan M, Mohd-Naim NF, Ahmed MU (2018) Trends and advances in electrochemiluminescence nanobiosensors. Sensors 18(1):166
Robert A, Liu Y, Nguyen M, Meunier B (2015) Regulation of copper and iron homeostasis by metal chelators: a possible chemotherapy for Alzheimer's disease. Acc Chem Res 48(5):1332–1339
Rocchitta G, Spanu A, Babudieri S, Latte G, Madeddu G, Galleri G, Manetti R (2016) Enzyme biosensors for biomedical applications: strategies for safeguarding analytical performances in biological fluids. Sensors 16(6):780
Rozand C (2014) Paper based analytical devices for point-of-care infectious disease testing. Eur J Clin Microbiol Infect Dis 33(2):147–156
Saadaoui M, Braiek M, Jaffrezic-Renault N, Raouafi N (2016) An ultrasensitive nanobiohybrid platform for glucose electrochemical biosensing based on ferrocenyl iminopropyl-modified silica nanoparticles. RSC Adv 6(52):46238–46243
Sabbagh MN, Lue LF, Fayard D, Shi J (2017) Increasing precision of clinical diagnosis of Alzheimer's disease using a combined algorithm incorporating clinical and novel biomarker data. Neurol Ther 6(1):83–95
Satvekar RK, Tiwale BM, Pawar SH (2014) Emerging trends in medical diagnosis: a thrust on nanotechnology. Med Chem 4:407–416
Saylan Y, Erdem Ö, Ünal S, Denizli A (2019) An alternative medical diagnosis method: biosensors for virus detection. Biosensors 9(2):65
Seven B, Bourourou M, Elouarzaki K, Constant JF, Gondran C, Holzinger M, Timur S (2013) Impedimetric biosensor for cancer cell detection. Electrochem Commun 37:36–39
Siegel RL, Miller KD, Jemal A (2019) Cancer statistics, 2019. CA-Cancer J Clin 69(1):7–34
Soleymani L, Li F (2017) Mechanistic challenges and advantages of biosensor miniaturization into the nanoscale. ACS Sens 2(4):458–467

Sun J, Xianyu Y, Jiang X (2014) Point-of-care biochemical assays using gold nanoparticle-implemented microfluidics. Chem Soc Rev 43(17):6239–6253

Tabatabaee RS, Golmohammadi H, Ahmadi SH (2019) Easy diagnosis of jaundice: a smartphone-based nanosensor bioplatform using photoluminescent bacterial nanopaper for point-of-care diagnosis of hyperbilirubinemia. ACS Sens 4(4):1063–1071

Tada H, Nohara A, Kawashiri MA (2018) Serum triglycerides and atherosclerotic cardiovascular disease: insights from clinical and genetic studies. Nutrients 10(11):1789

Thangamuthu M, Gabriel W, Santschi C, Martin O (2018) Electrochemical sensor for bilirubin detection using screen printed electrodes functionalized with carbon nanotubes and graphene. Sensors 18(3):800

Thévenot DR, Toth K, Durst RA, Wilson GS (2001) Electrochemical biosensors: recommended definitions and classification. Anal Lett 34(5):635–659

Tran S, DeGiovanni PJ, Piel B, Rai P (2017) Cancer nanomedicine: a review of recent success in drug delivery. Clin Transl Med 6(1):44

Tsai MT, Tarng DC (2019) Beyond a measure of liver function—bilirubin acts as a potential cardiovascular protector in chronic kidney disease patients. Int J Mol Sci 20(1):117

Van der Drift MA, Hol BE, Klaassen CH, Prinsen CF, van Aarssen YA, Donders R, Thunnissen FB (2010) Circulating DNA is a non-invasive prognostic factor for survival in non-small cell lung cancer. Lung Cancer 68(2):283–287

Vashistha R, Dangi AK, Kumar A, Chhabra D, Shukla P (2018) Futuristic biosensors for cardiac health care: an artificial intelligence approach. Biotech 8(8):358

Vidic J, Vizzini P, Manzano M, Kavanaugh D, Ramarao N, Zivkovic M, Gadjanski I (2019) Point-of-need DNA testing for detection of foodborne pathogenic bacteria. Sensors 19(5):1100

Vilela P, El-Sagheer A, Millar TM, Brown T, Muskens OL, Kanaras AG (2016) Graphene oxide-upconversion nanoparticle based optical sensors for targeted detection of mRNA biomarkers present in Alzheimer's disease and prostate cancer. ACS Sens 2(1):52–56

Viter R, Iatsunskyi I, Fedorenko V, Tumenas S, Balevicius Z, Ramanavicius A, Bechelany M (2016) Enhancement of electronic and optical properties of ZnO/Al2O3 nanolaminate coated electrospun nanofibers. J Phys Chem C 120(9):5124–5132

Walia S, Guliani A, Acharya A (2017) A theragnosis probe based on BSA/HSA-conjugated biocompatible fluorescent silicon nanomaterials for simultaneous in vitro cholesterol effluxing and cellular imaging of macrophage cells. ACS Sustain Chem Eng 5(2):1425–1435

Wang R, Di J, Ma J, Ma Z (2012) Highly sensitive detection of cancer cells by electrochemical impedance spectroscopy. Electrochim Acta 61:179–184

Wang L, Zhu F, Chen M, Xiong Y, Zhu Y, Xie S, Chen X (2019) Development of a "dual gates" locked, target-triggered nanodevice for point-of-care testing with a glucometer readout. ACS Sens 4(4):968–976

Wei H, Xu L, Ren J, Jia L (2012) Adsorption of bilirubin to magnetic multi-walled carbon nanotubes as a potential application in bound solute dialysis. Colloids Surf A Physicochem Eng Asp 405:38–44

Wu C, Liu X, Li Y, Du X, Wang X, Xu P (2014) Lipase-nanoporous gold biocomposite modified electrode for reliable detection of triglycerides. Biosens Bioelectron 53:26–30

Wu X, Luo L, Yang S, Ma X, Li Y, Dong C, Wu A (2015) Improved SERS nanoparticles for direct detection of circulating tumor cells in the blood. ACS Appl Mater Interfaces 7(18):9965–9971

Xie J, Jon S (2012) Magnetic nanoparticle-based theranostics. Theranostics 2(1):122

Yuan YR, Yuan R, Chai YQ, Zhuo Y, Miao XM (2009) Electrochemical amperometric immunoassay for carcinoembryonic antigen based on bi-layer nano-au and nickel hexacyanoferrates nanoparticles modified glassy carbon electrode. J Electroanal Chem 626(1–2):6–13

Zhang X, Liu S, Song X, Wang H, Wang J, Wang Y, Yu J (2019) Robust and universal SERS sensing platform for multiplexed detection of Alzheimer's disease core biomarkers using PAapt-AuNPs conjugates. ACS Sens 4(8):2140–2149

Zheng M, Ruan S, Liu S, Sun T, Qu D, Zhao H, Sun Z (2015) Self-targeting fluorescent carbon dots for diagnosis of brain cancer cells. ACS Nano 9(11):11455–11461

Zottel A, Videtic Paska A, Jovčevska I (2019) Nanotechnology meets oncology: nanomaterials in brain cancer research, diagnosis and therapy. Materials 12(10):1588

Zou Y, Zhang Y, Xu Y, Chen Y, Huang S, Lyu Y, Tan W (2018) Portable and label-free detection of blood bilirubin with graphene-isolated-Au-nanocrystals paper strip. Anal Chem 90 (22):13687–13694

Biomolecules Immobilized Nanomaterials and Their Biological Applications

5

Ashish K. Shukla, Mohini Verma, and Amitabha Acharya

Contents

A. K. Shukla · M. Verma · A. Acharya (✉)
Biotechnology Division, CSIR-Institute of Himalayan Bioresource Technology (CSIR-IHBT), Palampur, India

Academy of Scientific and Innovative Research (AcSIR), CSIR-Institute of Himalayan Bioresource Technology (CSIR-IHBT), Palampur, India
e-mail: amitabha@ihbt.res.in

A. Acharya (ed.), *Nanomaterial - Based Biomedical Applications in Molecular Imaging, Diagnostics and Therapy*, https://doi.org/10.1007/978-981-15-4280-0_5

Abstract

Biomolecules are the key factors for most of the biological processes. Immobilization is an impressive technique which is used to stabilize free biomolecules in solution. Immobilized biomolecules are more robust and more resistant to environmental changes. The main objective of the immobilization of biomolecules, namely., enzymes, is to enhance the economics of biocatalysis processes. Additionally, immobilization improves many properties of enzymes such as performance in organic solvents, pH tolerance, heat stability, and functional stability. More importantly, the heterogeneity of the immobilized biomolecule systems allows an easy recovery of both biomolecules and products, multiple reuse, continuous operation of enzymatic processes, rapid termination of reactions, and greater variety of bioreactor designs. This chapter presents an overview of the recent literature reports for biomolecule immobilization on nanomaterial surfaces by various techniques. It also demonstrates the need for immobilization and different applications at various platform.

Keywords

Biological applications · Biomolecules · Immobilization · Kinetics · Stability

Abbreviations

AMP	Antimicrobial peptide
AuNPs	Gold nanoparticles
CLE	Cross-linking enzymes
CLEAs	Cross-linked enzyme aggregations
dsDNA	Double-stranded deoxyribonucleic acid
EDC	Ethyl (dimethylaminopropyl) carbodiimide
FRET	Förster resonance energy transfer
HDAC	Histone deacetylase
MCH	6-mercato-1-hexanol
MNPs	Magnetic nanoparticles
MWCNTs	Multiwalled carbon nanotubes
NHS	N-hydroxysuccinimide
NMs	Nanomaterials
NPs	Nanoparticles
PEG	Polyethylene glycol
R & D	Research and development
SI-ATRP	Surface-initiated atom-transfer radical polymerization
ssDNA	Single-stranded deoxyribonucleic acid
SWCNTs	Single-walled carbon nanotubes

5.1 Introduction

Biomacromolecules, namely lipids, proteins, enzymes, and nucleic acids, are the most important systemic unit of cells. These show unique properties which majorly contribute as the center stage for different biochemical and physiological process of cells (Jia et al. 2018). With significant enhancement in R & D activities, biomacromolecules have been needed for various biomedical as well as industrial field. In the late 1920, the first enzyme immobilization (invertase) was reported by Nelson and Griffin on artificial aluminum hydroxide as a base via physical adsorption, which captured the interest of researchers to develop innovative strategies for immobilization (Zhang et al. 2015). Therefore, during 1920–1960, a lot of new and advantageous immobilization techniques were developed, for example, physical adsorption, covalent attachment, entrapment and encapsulation, and cross-linking attachment, etc., for the improvement of enzyme stability and productivity. Since 1960, these methods and processes were modified and applied at industrial and laboratory scale to improve the stability and activity of different biomolecules. But, few major challenges for using these molecules in artificial surroundings still remain, namely very limited use, problem in their upscaling, minimal recovery process, thermal and physiological instability, poor in vitro reaction ability, and poor binding efficiency toward substrate (Cao et al. 2003). Hence, there is an urgent need for alternative technologies to combat all these limitations. Product purification and repetitive usage of enzymes in long-term reaction followed by their recovery is also a big challenge for industrial applications. To overcome these obstacles, immobilization technique is the only preferred technology to economize the processed biomolecules on nanomaterial surface to improve stability, reusability, selectivity, and cost effectiveness (Ansari and Husain 2012; Brady and Jordaan 2009). Nanomaterials (NMs) act as a fruitful base or support for immobilization process, due to their various advantages such as large surface area to volume ratio, stabilizing biocatalysts efficiency, strong chemical as well as physical properties, etc. (Walia et al. 2019). Diffusion is a major problem in which there is high chances of substrate diffusion from reaction media. Additionally, many other obstacles are present during immobilization process, for example, false orientation and loss of catalytic activity of biomolecules (i.e., enzymes, proteins, ssDNA, etc.). Nowadays, nanoparticles are being used as a model system to combat the diffusion problems during immobilization process (Ansari and Husain 2012). Hence, the advantageous properties of NMs combine with innovative techniques would be an ideal approach. Here, in this chapter, we have described the advantages of different immobilization processes and their impact on related biological activities.

5.2 Methods of Immobilization and Their Advantages

The proper selection of immobilization method is the only key to overcome the limitation of biomolecules' unstability. In a specific reaction, it plays censorious role in evaluation of biomolecules activity and their characteristic features. The overall

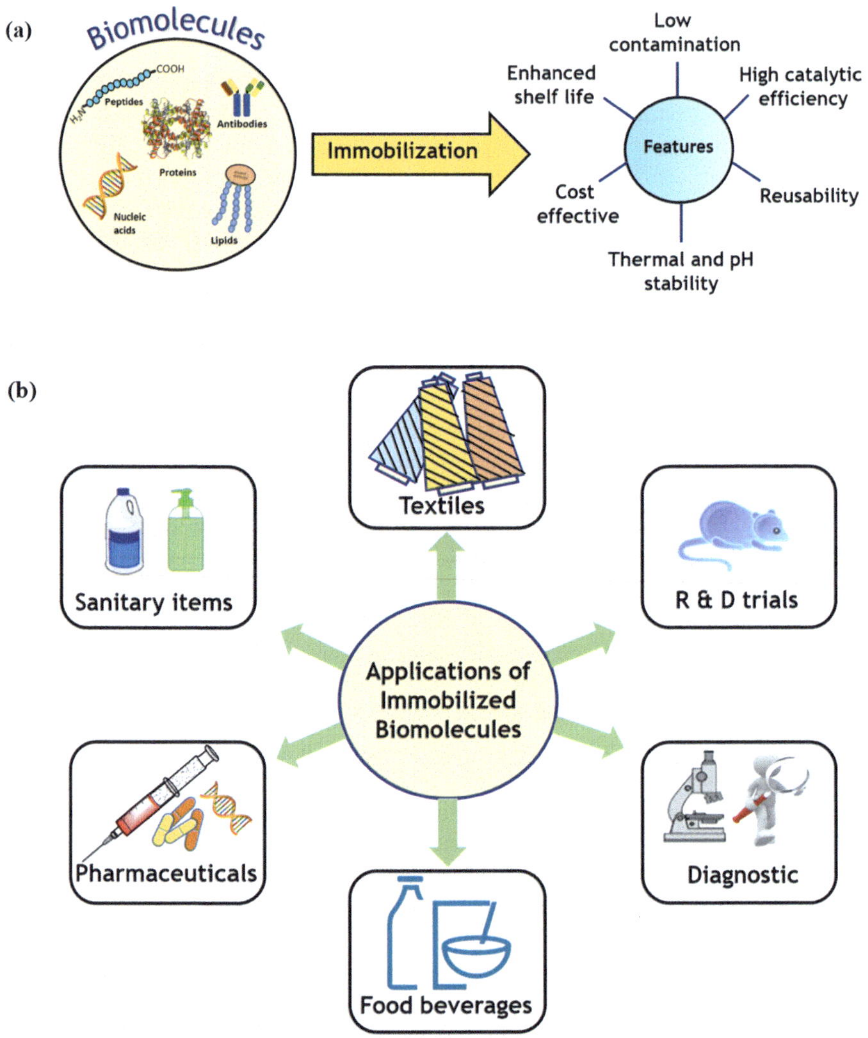

Fig. 5.1 (**a**) Schematic diagram showing different features of biomolecules immobilization and (**b**) their applications

system requires a better activity, performance, reusability, minimal cost of the process, and minimal toxicity. There are a lot of methods available for immobilization process, for example, covalent attachment, physical adsorption and entrapment, affinity tag approach, polymer cross-linking, etc. (Pundir 2015). Features of different immobilization techniques are shown in Fig. 5.1.

5.2.1 Covalent Attachment

Covalent attachment is a process in which biomolecules are attached to a specific insoluble or soluble base with the help of chemically active functional moiety. This technique is highly useful due to its robust advantage over biomolecule leakage and instability (Pundir 2015). Major advantages of this techniques are strong binding capability of substrate to biomolecule, high heat stability, flexibility of design, and wide range of modifications that can be done via click chemistry. Click chemistry does not only rule the era of covalent immobilization but also gives the strong perspective for rapid and stable immobilization of biomolecules on NMs surface (Nemzer et al. 2010). Various literature reports suggest that stimulation of NMs surface via active chemical moiety is the most important factor which gives direction for enzymes and protein immobilization (Secundo 2013). Similar to non-covalent immobilization, it can be used for unmodified biomolecules too, which only rely on natural functional moiety. For example, if proteins are exposed to any ester active support, they can readily react and form unstable conjugation in aqueous environment and undergo ester hydrolysis due to nucleophilicity of thiol (from cysteine) in aqueous media. To overcome this disadvantage, an aldehyde group can be coupled with amine group of proteins in the presence of ß-mercaptoethanol or sodium borohydride for the formation of stable secondary amide linkage (Homaei et al. 2013). Based on thiol reactivity toward biomolecules and metallic nanoparticles nowadays, chemisorption is widely used for covalent immobilization. For example, single strand DNA (ssDNA) was covalently attached on gold nanoparticles (AuNPs) via chemisorption process (Li et al. 2007; Yu et al. 2013). Template-mediated covalent attachment is also widely used for oligomers immobilization, that is, two different templates containing phenol and benzene are covalently attached with oligomers via copper-catalyzed azide–alkyne cycloaddition (Núñez-Villanueva et al. 2019). Biomolecules like enzymes and proteins are well known for their therapeutic activities such as cancer treatment, treatment of genetic disorder, as food and nutritional supplements, etc. Hence, it is most important to maintain the stability and native activity of biomolecules. PEGylation of L-asparaginase via covalent attachment reduces its immunogenicity and short half-life along with high maximum saturation velocity (V_{max}) and Michaelis constant (K_m) (Meneguetti et al. 2019). Venditti et al. immobilized lipase on hydrophilic AuNPs via covalent attachment with high residual activity and loading capability of ~95% and 65–70%, respectively (Venditti et al. 2015).

5.2.2 Physical Adsorption and Entrapment

Unlike covalent attachment, physical adsorption is a process in which biomolecules are adsorbed on NMs surface via electrostatic interaction, van der Waals forces, hydrogen bonding, etc. between immobilizing biomolecules and NMs (Li et al. 2007). It can minimize the deactivation of biomolecules which is often caused by covalent chemical modification during immobilization process. Nitrocellulose membranes or hydrophobic polymer–coated substrates are majorly involved in this technique (Sharma et al. 2014). The advantages of using this technique is that neither does it involve any reducing agents nor does it require any modified biomolecules. However, the major disadvantage of this technique is involvement of weak and reversible interactions as compared to covalent immobilization. Hence, due to this poor attachment ability, biomolecules could be leached out from substrate, which results in contamination and loss of activity of immobilized assembly (Wang et al. 2015a, b). Ionic hydrogels are widely used in this technique. Based on hydrophobic and hydrogen bond interactions, these immobilizes biomolecules in the pattern of polymer. This noncovalent immobilization of biomolecules is affected by pH, temperature, concentration, and physicochemical properties of reaction mixture, etc. (Cacicedo et al. 2019).

5.2.3 Cross-linking Attachment

A carrier-free attachment in which biomolecules are immobilized via bifunctional cross-linking agents is known as cross-linking immobilization, that is, cross-linking enzymes (CLE) (Cipolatti et al. 2014). This is a simple process for immobilization without using any support materials as compared to covalent and noncovalent immobilization techniques. So, these are mostly used in the form of sol–gel–sol, hydrogel, gelatinous matrix, etc. As per the literature reports, there are various strategies which produce stable, highly efficient, and low-cost immobilization of enzymes and proteins via cross-linking approach (Cacicedo et al. 2019; Cipolatti et al. 2014). The commonly used cross-linker is glutaraldehyde in which two end aldehyde groups freely react with amino group of biomolecules (Hartmann and Kostrov 2013). This approach results in gelatinous immobilization of biomolecules which cannot be specifically controlled due to improper aggregation of immobilized enzymes. There are various reports suggesting that cross-linked enzyme aggregations (CLEAs) physically aggregate enzymes in the aqueous solvent via water miscible solvents, salts, and non-ionic polymer addition (Rodrigues et al. 2013). Chitosan-coated magnetic nanomaterials have been recently used for cross-linking immobilization of enzymes with enhanced biological activity (Díaz-Hernández et al. 2018; Khoshnevisan et al. 2019).

5.2.4 Affinity Tagging

Affinity tagging is an innovative approach mainly used in protein–protein or protein–molecule interactions in a specific orientation (Hood et al. 2014). Biotin- and streptavidin-mediated conjugation of proteins with metallic NMs is widely explored in sensing applications. Inorganic NPs are widely chosen for bioaffinity interaction with various biomolecules namely proteins, peptides, ssDNA, and lectins along with few of the enzymes (Compostella et al. 2017). Bioaffinity interactions of protein–NPs have gained a lot of attention recently. This type of interaction influenced by physicochemical properties of NPs such as hydrophobicity, surface charge, shape, and size, etc. (Dar et al. 2019). Nowadays, proteins or peptide–NPs interaction are widely used in nanomedicine applications due to versatile properties of biological peptide-mediated stealth effect and biocompatibility. Hence, these nano assemblies work as a tool for targeted delivery of nanoparticles without immune system detection. Most of the NMs, that is mesoporous silica NPs, liposomes, polymeric NMs, AuNPs, magnetic NPs, etc., are used in protein–NPs–based nanomedicine development (Spicer et al. 2018). The different physicochemical properties which affect the immobilization and catalytic activity of the biomolecules have been discussed in the following section.

5.3 Physicochemical Properties of Biomolecules

Physicochemical properties are the most important parameters to optimize the effect of the surface on biomolecule conjugation. Here, we briefly discuss about the importance of thermodynamic process as aid to understand the biomolecules–NMs interaction. Thermodynamics of reaction determines how ligand and receptors act during any interaction. The standard thermodynamic equation which characterize the steps of biomolecules–NMs conjugates is as follows:

$$\Delta_r G_0^{agg} = \Delta_r G_0^{rel} + \Delta_r G_0^{bind} + \Delta_r G_0^{conj} \tag{5.1}$$

where, $\Delta_r G_0^{rel}$ is the standard Gibbs free energy change of released biomolecules from bioconjugates, $\Delta_r G_0^{bind}$ is binding of free biomolecules to the substrate, and $\Delta_r G_0^{conj}$ is the conjugation of substrate biomolecules to NMs (Rodriguez-Quijada et al. 2018).

5.3.1 Kinetic Parameters

Kinetic parameter describes the behavior of biomolecule binding with substrate. There are two well-known standard equations (Michaelis–Menten and Briggs–Haldane) which describe the dependency of biomolecule-catalyzed reaction rate on the substrate concentration via two parameters, namely, the catalytic constant (k_{cat}) and Michaelis–Menten constant (K_m). The maximum rate of reaction at saturated substrate concentration is determined via k_{cat} estimation, whereas, K_m is the substrate concentration at which the rate of reaction is half of the maximum velocity (V_{max}) of reaction. These two parameters can be easily characterized by Lineweaver–Burk plot via linear transformation (Choi et al. 2017). Rodrigues et al. explained the factors which affect the kinetic parameters of bioconjugates, namely, rigidification, effect of medium, effect of substrate or product partition, diffusional limitations, and mixture of biomolecules and their environment. An improper immobilization methodology produces strong rigidity in biomolecule structure due to multipoint covalent attachments, which facilitate structural deformities in conjugated biomolecules (Rodrigues et al. 2013). Generally, biomolecules are always immobilized in the presence of the substrate so that active sites are not hampered. In the presence of inactivating agents such as hydrogen peroxide, free radicals, dissolved gases, organic solvents, etc., immobilized biomolecules significantly affect the soluble environment. This facilitates disturbance in the hydrophilic–lipophilic balance of working mixture of bioconjugates. So to minimize or stabilize the medium effect, biomolecules were further modified using suitable polymers (Brady and Jordaan 2009).

5.3.2 pH and Temperature Dependence

The milieu of biomolecule immobilization plays an important role in maintaining the integrity of biomolecules on solid substrate. pH changes in reaction mixture affect the active confirmation of biomolecules immobilized on substrate. The microenvironment of the immobilized biomolecules and reaction mixture usually has unequal partitioning of H^+ and OH^- concentrations due to electrostatic interactions with the substrate, which often leads to the change in the pH activity profile. The pH effect on the activity of free and immobilized laccase was examined in the pH range 2.0–7.0 at 30 °C. The optimum pH for free laccase activity was found at pH 4.0 which was similar to that reported previously. On the other hand, the optimum pH for immobilized laccase activity was found to shift to pH 5.0 and the thermal stability also shifted to 60 °C (Mazlan and Hanifah 2017). Immobilization of laccase on nanofiber matrix broaden the pH range, which suggests that immobilization process protects enzyme activity in a wider pH range (Fatarella et al. 2014). Garg et al. examined the pH dependent immobilization of urease on glutathione-modified AuNPs. Immobilized urease retained higher enzymatic activity at pH 7.4 as compared to free one (Garg et al. 2015). Biomolecules are highly sensitive toward temperature. Higher temperature causes irreversible loss of catalytic activity. Talebi et al. showed that thermal stability of α-amylase was improved after their immobilization on nanocomposite (Talebi et al. 2016).

5.3.3 Redox Potential

Biomolecules such as oxidoreductase and metalloproteins have redox potential in their reaction milieu. One-third of the cellular process is involved with these types of biomolecules for electron transfer, that is, photosynthesis, respiration, metabolism, and molecular signaling. The functional properties of the oxidoreductases/metalloproteins are determined by the nature of their redox centers. Biomolecule engineering is a powerful approach that is used to incorporate biological and abiological redox cofactors with predictable structures and desirable functions for important biological and chemical applications (Prabhulkar et al. 2012). The immobilization of enzymes on electrodes is of significant interest in the development of enzymatic fuel cells (EFC) and biosensors. Siepenkoetter et al. reported immobilized glucose dehydrogenase on nanoporous AuNPs which showed strong oxidative properties compared to free enzyme (Siepenkoetter et al. 2017). Carbon nanomaterials show better electron transfer process, so nowadays these are directly engineered and used as nanozyme, which resembles the redox enzyme activity (Le Goff and Holzinger 2018). Zhou et al. developed reagentless glucose sensing via glucose oxidase immobilization on carbon nanotubes (CNTs) (Zhou et al. 2017).

5.4 Biomolecules Immobilized on Nanomaterials

5.4.1 Enzyme and Proteins

Enzymes are well-known biological catalysts and are involved in vital physiological functions of different living cells of animals, plants, and microorganisms. Nowadays, enzymes are well explored in many fields of applications, such as industrial food processing, biosensing, therapeutics, etc. Wang et al. reported a new way of alkaline phosphatase (ALP) immobilization on magnetic nanoparticles (MNP). They synthesized amine-functionalized silica-coated magnetic nanoparticles (Fe_3O_4@SiO_2@APES) and to this ALP was covalently attached with the help of sodium cyanoborohydride, which acts as reducing agent and form stable secondary amine linkage between enzyme and MNPs (Zhang et al. 2015). Likewise, lipase was covalently attached on MNP via carbodiimide-mediated direct binding between MNPs and immobilized enzyme and used for transesterification of fatty acid to biodiesel (Wang et al. 2013a, b). Li et al. reported immobilization process of glucose oxidase (GOx) on gold nanoparticle (AuNPs) via ethyl (dimethylaminopropyl) carbodiimide/N-hydroxysuccinimide (EDC/NHS) coupling reaction (Yu et al. 2013). Recently, cellulase was immobilized on MNP via cross-linking approach with high loading efficiency and increased reusability by ~50% (Ingle et al. 2017). Similarly, Lin et al. immobilized cellulase on chitosan-modified MNPs via carboxylmethyl cellulose as substrate for cross-linking and enhanced the reusability by ~84% (Lin et al. 2017). In the recent decades, immobilization process has increased the stability and activity of various therapeutic enzymes such as acetylcholinesterase, α-amylase, alcohol dehydrogenase, cytochrome P450, trypsin, peptides, proteins, etc., and these are mentioned in Table 5.1. Ganesana et al. reported the acetylcholinesterase immobilization on nickel NPs via site-specific 6-histidine affinity tagging (Ganesana et al. 2011). Similarly, Uygun et al. performed α-amylase immobilization on MNPs with enhanced activity and reusability by ~85% and ~10 times, respectively (Uygun et al. 2012). One of the important therapeutic enzyme L-asparaginase (L-ASNase) immobilization was reported by Monajati et al. They functionalized graphene oxide via aspartic acid and immobilized L-ASNase covalently with increased activity, recovery and reusability by 90%, 42%, and after 8 cycles, respectively (Monajati et al. 2018). Similarly, Baharman and Alemzadeh reported the immobilization of L-ASNase on calcium alginate beads via microencapsulation process and enhanced ~35% residual activity with ~46% loading efficiency (Bahraman and Alemzadeh 2017). Alcohol toxicity is a major concern

Table 5.1 Different types of biomacromolecules immobilized on nanomaterials (NMs) and their biological application

Name of biomacromolecules	NMs for immobilization	Immobilization techniques	Biological applications	References
Lipase B	Chitosan-modified magnetic NPs and organosilica NPs	Covalent attachment	Bio-based esterification of fatty acids	Hosseini et al. (2019) and Jiang et al. (2019a, b)
L-asparaginase	Aspartic acid–modified GO	Covalent attachment	Prevention of acrylamide formation in food industry products	Monajati et al. (2018)
Alcohol dehydrogenase	Titania nanoparticles	Physical adsorption	Substrate inhibition removal during formaldehyde and methanol reaction	Ghannadi et al. 2019
Urease	Zinc oxide NPs–modified gold electrode	Physical adsorption	Sensing of urea	Eghbali et al. (2015)
Streptokinase	Sol–gel–sol matrix of Boehmite NPs	Physical entrapment	Thombolytic application of blood clots	Chapurina et al. (2016)
Cytochrome P450	Ni^{2+}-functionalized MNPs	Affinity tagging and physical adsorption	Metabolism of fatty acid and hormones	Bahrami et al. (2017)
Pepsin and trypsin	Silica NPs gel and MNPs	Cross linking and covalent attachment	Proteolysis during food production process	Zhang et al. (2018) and Szałapata et al. (2016)
Phosphotriesterase	MNPs bioconjugates	Covalent attachment	Organophosphate degradation	Gebreyohannes et al. (2018)
t-plasminogen activator	Supermagnetic iron oxide NPs (SPIONs)	Covalent attachment	Fibrinolysis of plasma and thromboembolic disease	Heid et al. (2017)
B-galactosidase	Zinc oxide NPs	Covalent attachment	Degradation of lactose	Ansari et al. (2011)
Lysozyme	Chitosan nanofiber	Cross linking	Antimicrobial activity	Park et al. (2013)
Cholesterol oxidase	MNPs	Covalent attachment	Total cholesterol analysis in blood	Kouassi et al. (2005)
HA2 peptide	AuNPs	Covalent attachment	Endosomal escape	Cesbron et al. (2015)
Angiopep2	Polyethylene glycol liposome	Physical entrapment	Blood–brain barrier penetration	Endo-Takahashi et al. (2016)
Lactoferrin	Naflon-functionalized MWCNTs	Covalent attachment	Antimicrobial	Devi et al. (2017)

nowadays, therefore, Ghannadi et al. immobilized alcohol dehydrogenase (ADH) on titania nanoparticles via physical adsorption with increased residual activity and reusability by ~84% and 10 cycles, respectively (Ghannadi et al. 2019). Urea detection is vital part of pathological analysis for early detection of kidney-related malfunction. Therefore, Eghbali et al. reported urease immobilization on zinc oxide nanoparticles via physical adsorption for urea sensing at a pH range of 6–8 (Eghbali et al. 2015). Proteolytic proteins pepsin and trypsin were immobilized on silica NP gels and MNPs, respectively, with increased thermal stability (Szałapata et al. 2016; Zhang et al. 2018).

5.4.2 Nucleic Acids

Nucleic acid is a key molecule of whole living entity, and these molecules have very specific activities. Nowadays single-stranded DNA (ssDNA) are utilized as a versatile tool for biosensing application. The simplicity and advantage of these molecules allow us to design these for targeted sequence and transduce to form DNA hybrid for detection (Wang et al. 2015a, b). Compatible and better technique of DNA or RNA immobilization promotes efficient reactivity and orientation of immobilized molecules (Xu et al. 2015). Benvidi et al. developed the process for single-strand DNA (ssDNA) immobilization. They drop cast-thiolated ssDNA probe on gold electrode and incubate it for overnight reaction at room temperature followed by immerging in 6-mercapto-1-hexanol (MCH) to minimize the unspecific attachment (Benvidi et al. 2015; Thiruppathiraja et al. 2011). Zheng et al. reported polyaniline/graphene composite for electrostatic absorption DNA probe (Zheng et al. 2015). Similarly, Wang et al. explained the covalent immobilization of ssDNA via thiol and amine linkage with 3′ or 5′ end of sequence (Wang et al. 2011). Yang et al. immobilized aminated DNA probe on graphene oxide-fabricated single-walled carbon nanotube (SWCNTs) via EDC–NHS coupling reaction. Thus, the covalent immobilization is widely applicable approach in DNA probe immobilization on stable and compatible substrates (Yang et al. 2015). Similarly, Wang et al. reported DNA probe immobilization on glassy carbon electrode via electro-polymerization of Eriochrome Black T (sulphonic dye) (Wang et al. 2014).

5.4.3 Peptide and Lectins

Peptides are small sequence of amino acids, which are specifically designed for the biomedical applications such as antimicrobial activity, fluorometric or visual sensing of biological entity, tissue engineering, targeted delivery of NPs, etc. Similarly, antibodies are the immune-mediated biomolecules which are secreted from B-cells and play vital role in defense process (Spicer et al. 2018). Poon et al. performed two different peptides (cyclic RGD and myxoma virus derived) conjugation on AuNPs via secondary amine linkage between thiol and amine in the presence of sodium borohydride (Poon et al. 2015). Similarly, Guerrero et al. reported AuNPs and amphipathic peptide (CLPFFD) conjugation via simple covalent-bonding ring thiol conjugation (Guerrero et al. 2010). Qie et al. immobilized full-length CD47 peptide on three different sized carboxylated polystyrene NPs via carbodiimide-mediated click chemistry (Qie et al. 2016). Similarly, Schöttler et al. used polystyrene NPs for plasma protein conjugation and compared these nanoparticles toward the PEGylated counterpart (Schöttler et al. 2016). He et al. conjugated antimicrobial peptide (AMP) HHC36 on poly-SBMA–modified silica NPs via surface-initiated atom transfer radical polymerization (SI-ATRP) (He et al. 2018). Similarly, Sur et al. reported indolicidin AMP conjugation on carbon nanotube (CNT) and AuNPs via EDC–NHS conjugation (Sur et al. 2015). Chaudhary et al. conjugate three different AMP (TP359, 226, and 557) on silver-coated CNT via EDC–NHS coupling reaction (Chaudhari et al. 2019). Lectins are versatile carbohydrate-binding proteins having various applications namely cellular surface recognition, cellular permeability, cellular integrity, food production, etc. Sakuma et al. immobilized mucosal lectin (acetyl-D-galactosamine) on fluorescent polystyrene nanosphere via affinity tagging (Sakuma et al. 2009; Kitamura et al. 2015). Wang et al. studied the conjugation of glycans on fluorescent label silica NPs via photocoupling reaction (Wang et al. 2013a, b).

5.5 Biological Applications of Immobilized Biomolecules

5.5.1 Biosensing

Biosensing is an important application in current scenario for targeted analyte detection via converting biological signal into measurable units. Biosensing mainly includes electrochemical, thermal, colorimetry, and optical sensing. It involves few idealistic features, that is, specificity, reliability, cost effectiveness, and operational stability (Asal et al. 2019). Biosensing mechanism depends upon different biomolecules which are mainly used in this process, namely, enzymes, proteins, antibodies, DNA probes, lectins, peptides, etc. Immobilization and stabilization of these molecules on bioreceptor is the most important part of biosensing. Nowadays, nanoparticles are widely used in biosensing application via biomolecule immobilization because of their versatile nature and structural stability (Lee and Du-Duong 2018). In late 1960, first glucose biosensing was reported by Clark and Lyons based on glucose oxidase immobilization (Yoo and Lee 2010). Kim et al. reported portable urea sensing in biological sample by the help of urease immobilization on silk fibroin (Kim et al. 2018a, b). Visual ethanol biosensor was reported by Kuswandi et al. by alcohol oxidase immobilization on polyaniline membrane (Kuswandi et al. 2014). Thandavan et al. immobilized xanthine oxidase on MNPs via covalent interaction for xanthine sensing (Thandavan et al. 2013). Nesakumar et al. immobilized acetylcholinesterase on zinc oxide nanoparticles for carbosulphan biosensing (Nesakumar et al. 2016). Hassan et al. reported carrageenan biosensing via immobilization of calf thymus double-stranded DNA (dsDNA) on carbon-based screening electrode (Hassan et al. 2019). Surface plasmon resonance biosensing of targeted oligonucleotide via thiolated ssDNA probe immobilized on AuNPs was reported by Drozd et al. (Drozd et al. 2018). Ebrahimi et al. performed DNA-based biosensing for Ag^+ and Hg^{2+} ions in amalgam filling via covalent immobilization on modified multiwalled carbon nanotube (MWCNT) (Ebrahimi et al. 2018). He et al. reported electrochemical biosensor for silver deposition via specific prostate peptide and DNA probe conjugation on modified silica NPs (He et al. 2015).

5.5.2 Bioimaging

Bioimaging is a fluorescence-based imaging for tracking and deposition of synthesized NPs at targeting site. Fluorescent proteins and antibodies are widely used in this technique, namely, green fluorescent protein (GFP), red fluorescent protein (RFP), fluorescent tag secondary antibodies, etc. There are various developments in bioimaging techniques that have taken place in recent decades, but the precise tool and probe development for living cell imaging are still big challenging tasks. Skogs et al. clearly demonstrated the role of antibody in bioimaging process of endogenous proteins via GFP immobilization (Skogs et al. 2016). Miawaki et al. optimized long chain peptide–protein linker with enhanced FRET mediated bioimaging for targeted site (Miyawaki and Niino 2015). Fu et al. reported glycoprotein-mediated fluorescent nanoparticles synthesis which reduced the false and autofluorescence effect of live cells imaging (Fig. 5.2) (Fu et al. 2018).

5.5.3 Therapeutics

Therapeutics is a field of treatment, diagnostics, and prevention of disease site. Pharmacologically active biomolecules are widely used under therapeutics. Kouassi et al. immobilized cholesterol oxidase on MNPs for total cholesterol analysis present in serum for point-of-care solution (Kouassi et al. 2005). Similarly immobilization of lysozyme on chitosan nanofiber via cross-linkage for antimicrobial activity is another point-of-care application of immobilized biomolecule (Park et al. 2013). Lactose intolerance is one of the major problems in children where the indigestion of lactose causes severe stomach disturbance. Hence, Ansari et al. reported innovative approach for B-galactosidase immobilization on zinc oxide NPs via covalent binding for natural degradation of lactose (Ansari et al. 2011). Nanoparticles and protein interaction are very complex biophysical process due to which most of the NPs cannot properly act inside the body. Peptides are widely used biomolecules for specific targeted delivery of NPs and also these can bypass the protein corona formation due to their specificity toward the target. Cesbron et al. used HA2 biological peptide and immobilized on AuNPs via thiol conjugation, which improved the endosomal escape property of AuNPs in cellular environment (Cesbron et al. 2015). Brain is the most complex organ in human body, and it has lots of limitations toward the treatment related issues. Major obstacle of brain treatment is penetration of drug across blood–brain barrier (BBB). Endo-Takahashi

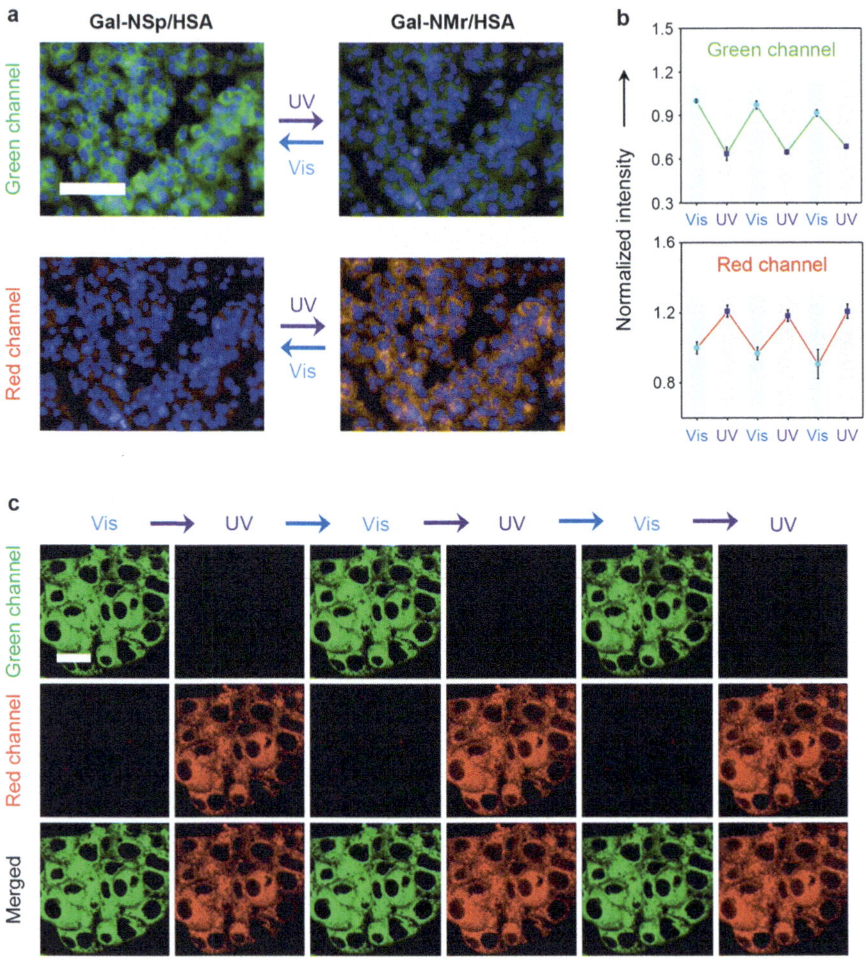

Fig. 5.2 (**a**)Fluorescent imaging of Hep-G2 cells with Gal-NSp/HAS (10 μM/100 μM) upon alternate UV/vis irradiation (green/red channel excitation: 460–490/520–550 nm; emission: 500–550/580–650 nm; scale bar = 100 μm). Hoechst 33342 were used to stain the cell nuclei. Error bars represent S. D. (n = 3). Intracellular fluorescence intensity switching between (**b**) Gal-NSp/HAS and Gal-NMr/HAS upon alternate UV/vis irradiation. (**c**) UV/vis cycling of the two imaging channels of Gal-NSp/HAS and Gal-NMr/HAS visualized by confocal laser-scanning microscopy (green/red channel excitation: 440/579 nm; emission: 535/603 nm; scale bar = 25 μm). (Adapted from Fu et al. 2018, reprint with permission)

et al. reported entrapment of angiopep-2 in liposomal nanocarrier which facilitates blood–brain barrier penetration of peptide conjugates (Endo-Takahashi et al. 2016). Similarly, Kale et al. reported dual mode application of liposome nanoparticle via immobilization of GFP plasmid and TATp peptide (Kale and Torchilin 2007).

5.5.4 Food Industry

Food production are the leading and foremost needed demand of market. Food products are packed and preserved via various biomolecules-mediated reaction, that is, potato fries are now treated with L-ASNase to minimize the content of acrylamide (Raveendran et al. 2018). Labrou et al. reported immobilization of L-ASNase on polyethylene glycol (PEG) for efficacy improvement of enzyme in food and clinical application (Labrou and Muharram 2016). Pasha et al. reported the utilization of pectinase in clarification of fruit juice (Pasha et al. 2013). Pectinase was immobilized on celite surface via adsorption to increase the stability and activity of the immobilized enzyme (Chauhan et al. 2015). Similarly, catalase was immobilized on magnetic-MWCNTs and has been used in food industry especially for hydrogen peroxide removal (Sirbu 2011; Rafiee et al. 2019). Peroxidase is widely used in food industry for nutritional and flavor quality development. Immobilization of peroxidase on porous silica increases the catalytic activity of enzyme, and it was used as catalytic reactor in industries (Regalado et al. 2004; Sahare et al. 2014). α-acetolactate dehydrogenase was immobilized on nanoflower/alginate microbeads and widely used in the beverage industry for shortening the time period of beer, liquor, and soft drink maturation process (Choi et al. 2015; Zhao et al. 2017). Lactose intolerance problems occur in many children and adults, and β-galactosidase deals with this issue. Hence, lactase (β-galactosidase) was immobilized on sodium alginate and l-carrageenan (1:5 ratio) beads and utilized as important enzyme in most of the milk food products (Souza et al. 2018). Similarly, phospholipase enzyme was immobilized on nanomagnetic carrier and widely used in breakdown of phospholipid to fatty acid, which is utilized in the production of vanaspati oil, ghee, etc. (Jiang et al. 2019a, b; Borrelli and Trono 2015).

5.5.5 Tissue Engineering

Tissue engineering is the advanced tool of biotechnology, which is the future of development of artificial organs, limbs, flesh, etc. In the recent decade, various nanoparticles are widely used for tissue engineering processes such as carbon nanomaterials, biodegradable polymeric nanomaterial, therapeutic enzyme immobilized nanocarrier, etc. Kim et al. reported that tunable hydrogel degradation facilitates the development of scaffolds for tissue regeneration. In this research, chitosan is cross-reacted with methacrylate to form hydrogels and the tunable degradation was occurred after lysozyme incorporation to hydrogels (Kim et al. 2018a, b). Sortase A is bacterial ligase protein which is recently explored for tissue engineering purpose. It was attached to biodegradable polymer via cross-linking in the presence of collagen which facilitates artificial tissue regeneration (Broguiere et al. 2018). Similarly, histone deacetylase (HDAC) is used in stem cell regeneration process. HDAC is an epigenetic factor which removes acetyl groups from histone so that chromatin assembly is tightly packed. This process facilitates stem cell for differentiation process (Lawlor and Yang 2019). Sivaraman et al. reported that

21 amino acid peptide, which was isolated from collagen, facilitate better adhesion property for HeLa cells and show significant lead in tissue regeneration (Sivaraman et al. 2019). Biomimetic peptides possess great potential for tissue regeneration. There are mainly three categories of peptides which can efficiently regenerate the tissue, namely, pro angiogenic peptide, anti-inflammatory peptide, and pro adherence peptide (Hosoyama et al. 2019).

5.6 Emerging Approach of Immobilized Biomolecules as Nanozyme

Nanomaterials with enzyme-mimicking properties are known as nanozymes. Nanozymes have effective catalytic sites for reactions which can serve as substitute of traditional enzymes for catalysis. Nanozyme possess inherent nanomaterial properties and can also act on a dynamic interfacing platform for complex biological environments (Zeng et al. 2018; Wu et al. 2019). Nowadays, nanomaterial engineering provides powerful biological interfaces to improve the performance of biosensors. Here, we have tried to compile few recent literature reports on biomolecules functionalized nano interfaces used for development of nanozyme. DNA, proteins, and peptides are utilized for functionalization of nanozyme. Different immobilization strategies for a range of DNA nanoscale biointerfaces, proteins recognitions, and peptide conjugations were optimized to modulate enzymatic and biological properties of the nanozyme system (Zhang et al. 2019). Fu et al. reported DNA-based platinum (Pt) nanomaterial for peroxidase mimic nanozyme application (Fu et al. 2014). Peptide-conjugated AuNPs were used as nanoprobe for quantification of integrin GPIIb/IIIa on the human erythroleukemia cell line. In addition, this nanoprobe also mimicked peroxidase activity which increased the sensitivity of integrin quantification on colorimetric basis (Gao et al. 2015). Similarly, anti-*E. coli* antibody was conjugated on platinum (Pt)–coated gold NPs for sensitive detection of bacteria via lateral flow immunoassay (Bradbury et al. 2019).

5.7 Conclusion and Future Aspects

Immobilization of biomolecules is a powerful tool for the development of novel biologically active nanomaterials. It is widely applicable for low-cost production and reusability of biomolecules. Within a few decades, scientific community and industries are witnessing remarkable achievements in immobilization techniques. The versatility of biomolecules plays an important role in the field of controlled utilization and modulation of these molecules with nanoparticles. Nanotechnology is providing solutions to all the limitations related to NPs–biomolecules conjugates. As we have seen in this chapter, many biomolecules can act in a similar manner under different immobilization conditions. Biological application of NPs and their conjugates are widely feasible due to enhanced stability and specificity. The rapid development in nanotechnology also makes the preparation of various nanoparticles more affordable. Biomolecules–NPs conjugates (bioconjugates) not only possess therapeutic and diagnostic activity but also inherit some intrinsic properties of nanoparticles, which benefit biosensing, diagnostics, cell killing, healing, and many other areas in the biomedical field. However, apart from impressive literature reports, several disadvantages of bioconjugates such as aggregation, precipitation, and bio-incompatibility are still challenging concerns. Despite some limitations, in immobilization process, biomolecules are still utilized for high-end applications. In the recent decade, tissue engineering application is the most attractive application in which carbon nanomaterials and polymeric nanomaterials are currently well explored. Recent literatures also suggest that biomolecules conjugated with nanomaterials can be used as surrogated artificial enzymes known as nanozymes. Thus, it can be commented that immobilization of biomolecules on the surface of advance substrates such as nanomaterials will pave the path for future applied science and technological outcomes.

Acknowledgment The authors would like to thank Director, Council of Scientific and Industrial Research–Institute of Himalayan Bioresource Technology (CSIR-IHBT) for his constant support and encouragement. AA acknowledges financial assistance in the form of project grant MLP-0201 from CSIR and GAP-0214 (EMR/2016/003027) from Department of Science & Technology (DST), Government of India. AKS and MV acknowledge CSIR-GATE, and DST, Government of India for their respective fellowships. The CSIR-IHBT communication number of this manuscript is 4504.

References

Ansari SA, Husain Q (2012) Potential applications of enzymes immobilized on/in nano materials: a review. Biotechnol Adv 30(3):512–523

Ansari SA, Husain Q, Qayyum S, Azam A (2011) Designing and surface modification of zinc oxide nanoparticles for biomedical applications. Food Chem Toxicol 49(9):2107–2115

Asal M, Özen Ö, Şahinler M, Baysal HT, Polatoğlu İ (2019) An overview of biomolecules, immobilization methods and support materials of biosensors. Sens Rev 39(3):377–386

Bahraman F, Alemzadeh I (2017) Optimization of l-asparaginase immobilization onto calcium alginate beads. Chem Eng Commun 204(2):216–220

Bahrami A, Vincent T, Garnier A, Larachi F, Boukouvalas J, Iliuta MC (2017) Noncovalent immobilization of optimized bacterial cytochrome P450 BM3 on functionalized magnetic nanoparticles. Ind Eng Chem Res 56(39):10981–10989

Benvidi A, Firouzabadi AD, Tezerjani MD, Moshtaghiun SM, Mazloum-Ardakani M, Ansarin A (2015) A highly sensitive and selective electrochemical DNA biosensor to diagnose breast cancer. J Electroanal Chem 750:57–64

Borrelli GM, Trono D (2015) Recombinant lipases and phospholipases and their use as biocatalysts for industrial applications. Int J Mol Sci 16(9):20774–20840

Bradbury DW, Azimi M, Diaz AJ, Pan AA, Falktoft CH, Wu BM, Kamei DT (2019) Automation of biomarker preconcentration, capture, and Nanozyme signal enhancement on paper-based devices. Anal Chem 91(18):12046–12054

Brady D, Jordaan J (2009) Advances in enzyme immobilisation. Biotechnol Lett 31(11):1639–1650

Broguiere N, Formica F, Barreto G, Zenobi-Wong M (2018) Sortase A as a cross-linking enzyme in tissue engineering. Acta Biomater 77:182–190

Cacicedo ML, Manzo RM, Municoy S, Bonazza HL, Islan GA, Desimone M, Castro GR (2019) Immobilized enzymes and their applications. In: Advances in enzyme technology. Elsevier, Amsterdam, pp 169–200

Cao L, Langen LV, Sheldon RA (2003) Immobilised enzymes: carrier bound or carrier-free? Curr Opin Biotechnol 14(4):387–394

Cesbron Y, Shaheen U, Free P, Lévy R (2015) TAT and HA2 facilitate cellular uptake of gold nanoparticles but do not lead to cytosolic localisation. PLoS One 10(4):e0121683

Chapurina YE, Drozdov AS, Popov I, Vinogradov VV, Dudanov IP (2016) Streptokinase@ alumina nanoparticles as a promising thrombolytic colloid with prolonged action. J Mater Chem B 4(35):5921–5928

Chaudhari AA, Joshi S, Vig K, Sahu R, Dixit S, Baganizi R, Pillai S (2019) A three-dimensional human skin model to evaluate the inhibition of Staphylococcus aureus by antimicrobial peptide-functionalized silver carbon nanotubes. J Biomater Appl 33(7):924–934

Chauhan S, Vohra A, Lakhanpal A, Gupta R (2015) Immobilization of commercial pectinase (Polygalacturonase) on celite and its application in juice clarification. J Food Process Preserv 39(6):2135–2141

Choi EJ, Ahn HW, Kim WJ (2015) Effect of α-acetolactate decarboxylase on diacetyl content of beer. Food Sci Biotechnol 24(4):1373–1380

Choi B, Rempala GA, Kim JK (2017) Beyond the Michaelis-Menten equation: accurate and efficient estimation of enzyme kinetic parameters. Sci Rep 7(1):17018

Cipolatti EP, Silva MJA, Klein M, Feddern V, Feltes MMC, Oliveira JV, de Oliveira D (2014) Current status and trends in enzymatic nanoimmobilization. J Mol Catal B Enzym 99:56–67

Compostella F, Pitirollo O, Silvestri A, Polito L (2017) Glyco-gold nanoparticles: synthesis and applications. Beilstein J Org Chem 13(1):1008–1021

Dar AI, Walia S, Acharya A (2019) Molecular recognition based rapid diagnosis of immunoglobulins via proteomic profiling of protein-nanoparticle complexes. Int J Biol Macromol 138:156–167

Devi KS, Mahalakshmi VT, Ghosh AR, Kumar AS (2017) Unexpected co-immobilization of lactoferrin and methylene blue from milk solution on a Nafion/MWCNT modified electrode and application to hydrogen peroxide and lactoferrin biosensing. Electrochim Acta 244:26–37

Díaz-Hernández A, Gracida J, García-Almendárez BE, Regalado C, Núñez R, Amaro-Reyes A (2018) Characterization of magnetic nanoparticles coated with chitosan: a potential approach for enzyme immobilization. J Nanomater 9468574:11

Drozd M, Pietrzak MD, Malinowska E (2018) SPRi-based biosensing platforms for detection of specific DNA sequences using thiolate and dithiocarbamate assemblies. Front Chem 6:173

Ebrahimi M, Raoof JB, Ojani R (2018) Sensitive electrochemical DNA-based biosensors for the determination of Ag^+ and Hg^{2+} ions and their application in analysis of amalgam filling. J Iran Chem Soc 15(8):1871–1880

Eghbali M, Farahbakhsh A, Rohani A, Pour AN (2015) Urea biosensor based on immobilization of urease on ZnO nanoparticles. Orient J Chem 31(2):1237–1242

Endo-Takahashi Y, Ooaku K, Ishida K, Suzuki R, Maruyama K, Negishi Y (2016) Preparation of angiopep-2 peptide-modified bubble liposomes for delivery to the brain. Biol Pharm Bull 39 (6):977–983

Fatarella E, Spinelli D, Ruzzante M, Pogni R (2014) Nylon 6 film and nanofiber carriers: preparation and laccase immobilization performance. J Mol Catal B Enzym 102:41–47

Fu J, Reinhold J, Woodbury NW (2011) Peptide-modified surfaces for enzyme immobilization. PLoS One 6(4):e18692

Fu Y, Zhao X, Zhang J, Li W (2014) DNA-based platinum nanozymes for peroxidase mimetics. J Phys Chem C 118(31):18116–18125

Fu Y, Han HH, Zhang J, He XP, Feringa BL, Tian H (2018) Photocontrolled fluorescence "double-check" bioimaging enabled by a glycoprobe–protein hybrid. J Am Chem Soc 140 (28):8671–8674

Ganesana M, Istarnboulie G, Marty J-L, Noguer T, Andreescu S (2011) Site-specific immobilization of a (His) 6-tagged acetylcholinesterase on nickel nanoparticles for highly sensitive toxicity biosensors. Biosens Bioelectron 30:43–48
Gao L, Liu M, Ma G, Wang Y, Zhao L, Yuan Q, Zhao Y (2015) Peptide-conjugated gold nanoprobe: intrinsic nanozyme-linked immunsorbant assay of integrin expression level on cell membrane. ACS Nano 9(11):10979–10990
Garg S, De A, Mozumdar S (2015) pH-dependent immobilization of urease on glutathione-capped gold nanoparticles. J Biomed Mater Res Part A 103(5):1771–1783
Gebreyohannes AY, Mazzei R, Marei Abdelrahim MY, Vitola G, Porzio E, Manco G, Giorno L (2018) Phosphotriesterase-magnetic nanoparticle bioconjugates with improved enzyme activity in a biocatalytic membrane reactor. Bioconjug Chem 29(6):2001–2008
Ghannadi S, Abdizadeh H, Miroliaei M, Saboury AA (2019) Immobilization of alcohol dehydrogenase on titania nanoparticles to enhance enzyme stability and remove substrate inhibition in the reaction of formaldehyde to methanol. Ind Eng Chem Res 58(23):9844–9854
Guerrero S, Araya E, Fiedler JL, Arias JI, Adura C, Albericio F, Kogan MJ (2010) Improving the brain delivery of gold nanoparticles by conjugation with an amphipathic peptide. Nanomedicine 5(6):897–913
Hartmann M, Kostrov X (2013) Immobilization of enzymes on porous silica–benefits and challenges. Chem Soc Rev 42(15):6277–6289
Hassan RA, Heng LY, Tan LL (2019) Novel DNA biosensor for direct determination of carrageenan. Sci Rep 9(1):6379
He Y, Xie S, Yang X, Yuan R, Chai Y (2015) Electrochemical peptide biosensor based on in situ silver deposition for detection of prostate specific antigen. ACS Appl Mater Interfaces 7 (24):13360–13366
He J, Chen J, Hu G, Wang L, Zheng J, Zhan J, Wang Y (2018) Immobilization of an antimicrobial peptide on silicon surface with stable activity by click chemistry. J Mater Chem B 6(1):68–74
Heid S, Unterweger H, Tietze R, Friedrich R, Weigel B, Cicha I, Lyer S (2017) Synthesis and characterization of tissue plasminogen activator—functionalized superparamagnetic iron oxide nanoparticles for targeted fibrin clot dissolution. Int J of Mol Sci 18(9):1837
Homaei AA, Sariri R, Vianello F, Stevanato R (2013) Enzyme immobilization: an update. J Chem Bio 6(4):185–205
Hood ED, Chorny M, Greineder CF, Alferiev IS, Levy RJ, Muzykantov VR (2014) Endothelial targeting of nanocarriers loaded with antioxidant enzymes for protection against vascular oxidative stress and inflammation. Biomaterials 35(11):3708–3715
Hosoyama K, Lazurko C, Muñoz M, McTiernan CD, Alarcon EI (2019) Peptide-based functional biomaterials for soft-tissue repair. Front Bioeng Biotechnol 7:205
Hosseini SM, Kim SM, Sayed M, Younesi H, Bahramifar N, Park JH, Pyo SH (2019) Lipase-immobilized chitosan-crosslinked magnetic nanoparticle as a biocatalyst for ring opening esterification of itaconic anhydride. Biochem Eng J 143:141–150
Ingle AP, Rathod J, Pandit R, da Silva SS, Rai M (2017) Comparative evaluation of free and immobilized cellulase for enzymatic hydrolysis of lignocellulosic biomass for sustainable bioethanol production. Cellulose 24(12):5529–5540
Jia M, Li S, Zang L, Lu X, Zhang H (2018) Analysis of biomolecules based on the surface enhanced Raman spectroscopy. Nano 8(9):730
Jiang Y, Du J, Tang H, Zhang X, Li W, Wang L, Yu D (2019a) Synthesis and application of nanomagnetic immobilized phospholipase C. J Chem 2019:5951793
Jiang Y, Liu H, Wang L, Zhou L, Huang Z, Ma L, He Y, Shi L, Gao J (2019b) Virus-like organosilica nanoparticles for lipase immobilization: characterization and biocatalytic applications. Biochem Eng J 144:125–134
Kale AA, Torchilin VP (2007) Enhanced transfection of tumor cells in vivo using "smart" pH-sensitive TAT-modified pegylated liposomes. J Drug Target 15(7–8):538–545
Khoshnevisan K, Poorakbar E, Baharifar H, Barkhi M (2019) Recent advances of cellulase immobilization onto magnetic nanoparticles: an update review. Magnetochemistry 5(2):36

Kim S, Cui ZK, Koo B, Zheng J, Aghaloo T, Lee M (2018a) Chitosan–lysozyme conjugates for enzyme-triggered hydrogel degradation in tissue engineering applications. ACS Appl Mater Interfaces 10(48):41138–41145
Kim K, Lee J, Moon B, Seo Y, Park C, Park M, Sung G (2018b) Fabrication of a urea biosensor for real-time dynamic fluid measurement. Sensors 18(8):2607
Kitamura T, Sakuma S, Shimosato M, Higashino H, Masaoka Y, Kataoka M, Koike S (2015) Specificity of lectin-immobilized fluorescent nanospheres for colorectal tumors in a mouse model which better resembles the clinical disease. Contrast Media Mol Imaging 10(2):135–143
Kouassi GK, Irudayaraj J, McCarty G (2005) Examination of cholesterol oxidase attachment to magnetic nanoparticles. J Nanobiotechnol 3(1):1
Kuswandi B, Irmawati T, Hidayat M, Ahmad M (2014) A simple visual ethanol biosensor based on alcohol oxidase immobilized onto polyaniline film for halal verification of fermented beverage samples. Sensors 14(2):2135–2149
Labrou NE, Muharram MM (2016) Biochemical characterization and immobilization of Erwinia carotovora L-asparaginase in a microplate for high-throughput biosensing of L-asparagine. Enzyme Microb Technol 92:86–93
Lawlor L, Yang XB (2019) Harnessing the HDAC–histone deacetylase enzymes, inhibitors and how these can be utilised in tissue engineering. Int J Oral Sci 11(2):20
Le Goff A, Holzinger M (2018) Molecular engineering of the bio/nano-interface for enzymatic electrocatalysis in fuel cells. Sustain Energy Fuels 2(12):2555–2566
Lee CK, Au-Duong AN (2018) Enzyme immobilization on nanoparticles: recent applications. Emerg Areas Bioeng 1:67–80
Li D, He Q, Cui Y, Duan L, Li J (2007) Immobilization of glucose oxidase onto gold nanoparticles with enhanced thermostability. Biochem Biophys Res Commun 355(2):488–493
Lin Y, Liu X, Xing Z, Geng Y, Wilson J, Wu D, Kong H (2017) Preparation and characterization of magnetic Fe_3O_4–chitosan nanoparticles for cellulase immobilization. Cellulose 24 (12):5541–5550
Mazlan SZ, Hanifah SA (2017) Effects of temperature and pH on immobilized laccase activity in conjugated methacrylate-acrylate microspheres. Int J Poly Sci 2017:5657271
Meneguetti GP, Santos JHPM, Obreque KMT, Barbosa CMV, Monteiro G, Farsky SHP, Pessoa-Junior A (2019) Novel site-specific PEGylated L-asparaginase. PLoS One 14(2):0211951
Miyawaki A, Niino Y (2015) Molecular spies for bioimaging—fluorescent protein-based probes. Mol Cell 58(4):632–643
Mogharabi M, Nassiri-Koopaei N, Bozorgi-Koushalshahi M, Nafissi-Varcheh N, Bagherzadeh G, Faramarzi MA (2012) Immobilization of laccase in alginate-gelatin mixed gel and decolorization of synthetic dyes. Bioinorg Chem Appl 2012:823830
Monajati M, Borandeh S, Hesami A, Mansouri D, Tamaddon AM (2018) Immobilization of L-asparaginase on aspartic acid functionalized graphene oxide nanosheet: enzyme kinetics and stability studies. Chem Eng J 354:1153–1163
Nemzer LR, Schwartz A, Epstein AJ (2010) Enzyme entrapment in reprecipitated polyaniline nano- and microparticles. Macromolecules 43(9):4324–4330
Nesakumar N, Sethuraman S, Krishnan UM, Rayappan JBB (2016) Electrochemical acetylcholinesterase biosensor based on ZnO nanocuboids modified platinum electrode for the detection of carbosulfan in rice. Biosens Bioelectron 77:1070–1077
Núñez-Villanueva D, Ciaccia M, Iadevaia G, Sanna E, Hunter CA (2019) Sequence information transfer using covalent template-directed synthesis. Chem Sci 10(20):5258–5266
Park JM, Kim M, Park HS, Jang A, Min J, Kim YH (2013) Immobilization of lysozyme-CLEA onto electrospun chitosan nanofiber for effective antibacterial applications. Int J Bio Macromol 54:37–43
Pasha KM, Anuradha P, Subbarao D (2013) Applications of pectinases in industrial sector. Int J Pure Appl Sci Technol 16(1):89
Poon W, Zhang X, Bekah D, Teodoro JG, Nadeau JL (2015) Targeting B16 tumors in vivo with peptide-conjugated gold nanoparticles. Nanotechnology 26(28):285101

Prabhulkar S, Tian H, Wang X, Zhu JJ, Li CZ (2012) Engineered proteins: redox properties and their applications. Antioxid Redox Signal 17(12):1796–1822
Pundir CS (ed) (2015) Enzyme noparticles, vol 69. William Andrew Publishing, Boston
Qie Y, Yuan H, Von Roemeling CA, Chen Y, Liu X, Shih KD, Kim BY (2016) Surface modification of nanoparticles enables selective evasion of phagocytic clearance by distinct macrophage phenotypes. Sci Rep 6:26269
Rafiee-Pour HA, Nejadhosseinian M, Firouzi M, Masoum S (2019) Catalase immobilization onto magnetic multi-walled carbon nanotubes: optimization of crucial parameters using response surface methodology. New J Chem 43(2):593–600
Raveendran S, Parameswaran B, Beevi Ummalyma S, Abraham A, Kuruvilla Mathew A, Madhavan A, Pandey A (2018) Applications of microbial enzymes in food industry. Food Technol Biotechnol 56(1):16–30
Regalado C, García-Almendárez BE, Duarte-Vázquez MA (2004) Biotechnological applications of peroxidases. Phytochem Rev 3(1–2):243–256
Rodrigues RC, Ortiz C, Berenguer-Murcia Á, Torres R, Fernández-Lafuente R (2013) Modifying enzyme activity and selectivity by immobilization. Chem Soc Rev 42(15):6290–6307
Rodriguez-Quijada C, Sánchez-Purrà M, de Puig H, Hamad-Schifferli K (2018) Physical properties of biomolecules at the nanomaterial interface. J Phys Chem B 122(11):2827–2840
Sahare P, Ayala M, Vazquez-Duhalt R, Agrawal V (2014) Immobilization of peroxidase enzyme onto the porous silicon structure for enhancing its activity and stability. Nanoscale Res Lett 9 (1):409
Sakuma S, Yano T, Masaoka Y, Kataoka M, Hiwatari KI, Tachikawa H, Tang L (2009) In vitro/ in vivo biorecognition of lectin-immobilized fluorescent nanospheres for human colorectal cancer cells. J Control Release 134(1):2–10
Schöttler S, Becker G, Winzen S, Steinbach T, Mohr K, Landfester K, Wurm FR (2016) Protein adsorption is required for stealth effect of poly (ethylene glycol)-and poly (phosphoester)-coated nanocarriers. Nat Nanotechnol 11(4):372
Secundo F (2013) Conformational changes of enzymes upon immobilisation. Chem Soc Rev 42 (15):6250–6261
Sharma B, Mandani S, Sarma TK (2014) Enzymes as bionanoreactors: glucose oxidase for the synthesis of catalytic Au nanoparticles and Au nanoparticle–polyaniline nanocomposites. J Mater Chem B 2(26):4072–4079
Siepenkoetter T, Salaj-Kosla U, Xiao X, Conghaile PÓ, Pita M, Ludwig R, Magner E (2017) Immobilization of redox enzymes on nanoporous gold electrodes: applications in biofuel cells. Chem Plus Chem 82(4):553–560
Sirbu T (2011) The searching of active catalase producers among the microscopic fungi. An Univ Oradea Fasc Biol 18(2):164–167
Sivaraman K, Muthukumar K, Shanthi C (2019) A potential bioactive peptide candidate for biomaterial and tissue engineering applications. Life Sci 226:140–148
Skogs M, Stadler C, Schutten R, Hjelmare M, Gnann C, Björk L, Lundberg E (2016) Antibody validation in bioimaging applications based on endogenous expression of tagged proteins. J Proteome Res 16(1):147–155
Souza CJ, Garcia-Rojas EE, Favaro-Trindade CS (2018) Lactase (β-galactosidase) immobilization by complex formation: impact of biopolymers on enzyme activity. Food Hydrocoll 83:88–96
Spicer CD, Jumeaux C, Gupta B, Stevens MM (2018) Peptide and protein nanoparticle conjugates: versatile platforms for biomedical applications. Chem Soc Rev 47(10):3574–3620
Sur A, Pradhan B, Banerjee A, Aich P (2015) Immune activation efficacy of indolicidin is enhanced upon conjugation with carbon nanotubes and gold nanoparticles. PLoS One 10(4):e0123905
Szałapata K, Osińska-Jaroszuk M, Bryjak J, Jaszek M, Jarosz-Wilkołazka A (2016) Novel application of porous and cellular materials for covalent immobilization of pepsin. Braz J Chem Eng 33 (2):251–260
Talebi M, Vaezifar S, Jafary F, Fazilati M, Motamedi S (2016) Stability improvement of immobilized α-amylase using nano pore zeolite. Iran J Biotechnol 14(1):33

Thandavan K, Gandhi S, Sethuraman S, Rayappa JBB, Krishnan UM (2013) Development of electrochemical biosensor with nano-interface for xanthine sensing–a novel approach for fish freshness estimation. Food Chem 139(1-4):963–969

Thiruppathiraja C, Kamatchiammal S, Adaikkappan P, Santhosh DJ, Alagar M (2011) Specific detection of Mycobacterium sp. genomic DNA using dual labeled gold nanoparticle based electrochemical biosensor. Anal Biochem 417(1):73–79

Uygun DA, Öztürk N, Akgöl S, Denizli A (2012) Novel magnetic nanoparticles for the hydrolysis of starch with Bacillus licheniformis α-amylase. J Appl Poly Sci 123(5):2574–2581

Venditti I, Palocci C, Chronopoulou L, Fratoddi I, Fontana L, Diociaiuti M, Russo MV (2015) Candida rugosa lipase immobilization on hydrophilic charged gold nanoparticles as promising biocatalysts: activity and stability investigations. Colloids Surf B Biointerfaces 131:93–101

Walia S, Shukla AK, Sharma C, Acharya A (2019) Engineered bright blue-and red-emitting carbon dots facilitate synchronous imaging and inhibition of bacterial and cancer cell progression via $^1O_2$-mediated DNA damage under photoirradiation. ACS Biomater Sci Eng 5(4):1987–2000

Wang Q, Zhang B, Lin X, Weng W (2011) Hybridization biosensor based on the covalent immobilization of probe DNA on chitosan–mutiwalled carbon nanotubes nanocomposite by using glutaraldehyde as an arm linker. Sens Actuators B Chem 156(2):599–605

Wang X, Matei E, Deng L, Koharudin L, Gronenborn AM, Ramström O, Yan M (2013a) Sensing lectin–glycan interactions using lectin super-microarrays and glycans labeled with dye-doped silica nanoparticles. Biosens Bioelectron 47:258–264

Wang S, Su P, Huang J, Wu J, Yang Y (2013b) Magnetic nanoparticles coated with immobilized alkaline phosphatase for enzymolysis and enzyme inhibition assays. J Mater Chem B 1 (12):1749–1754

Wang L, Liao X, Ding Y, Gao F, Wang Q (2014) DNA biosensor based on a glassy carbon electrode modified with electropolymerized Eriochrome Black T. Microchim Acta 181 (1-2):155–162

Wang XY, Jiang XP, Li Y, Zeng S, Zhang YW (2015a) Preparation Fe_3O_4@ chitosan magnetic particles for covalent immobilization of lipase from Thermomyces lanuginosus. Int J Bio Macromol 75:44–50

Wang K, Lei Y, Zhong GX, Zheng YJ, Sun ZL, Peng HP, Lin XH (2015b) Dual-probe electrochemical DNA biosensor based on the "Y" junction structure and restriction endonuclease assisted cyclic enzymatic amplification for detection of double-strand DNA of PML/RARα related fusion gene. Biosens Bioelectron 71:463–469

Wu J, Wang X, Wang Q, Lou Z, Li S, Zhu Y, Wei H (2019) Nanomaterials with enzyme-like characteristics (nanozymes): next-generation artificial enzymes (II). Chem Soc Rev 48 (4):1004–1076

Xu B, Zheng D, Qiu W, Gao F, Jiang S, Wang Q (2015) An ultrasensitive DNA biosensor based on covalent immobilization of probe DNA on fern leaf-like α- Fe_2O_3 and chitosan Hybrid film using terephthalaldehyde as arm-linker. Biosens Bioelectron 72:175–181

Yang L, Li X, Yan S, Wang M, Liu P, Dong Y, Zhang C (2015) Single-walled carbon nanotubes–carboxyl-functionalized graphene oxide-based electrochemical DNA biosensor for thermolabile hemolysin gene detection. Anal Methods 7(12):5303–5310

Yoo EH, Lee SY (2010) Glucose biosensors: an overview of use in clinical practice. Sensors 10 (5):4558–4576

Yu CY, Huang LY, Kuan I, Lee SL (2013) Optimized production of biodiesel from waste cooking oil by lipase immobilized on magnetic nanoparticles. Int J Mol Sci 14(12):24074–24086

Zeng C, Lu N, Wen Y, Liu G, Zhang R, Zhang J, Zhang M (2018) Engineering nanozymes using DNA for catalytic regulation. ACS Appl Mater Interfaces 11(2):1790–1799

Zhang Y, Ge J, Liu Z (2015) Enhanced activity of immobilized or chemically modified enzymes. ACS Catal 5(8):4503–4513

Zhang L, Wang B, Wang S, Zhang W (2018) Recyclable trypsin immobilized magnetic nanoparticles based on hydrophilic polyethylenimine modification and the proteolytic characteristics. Anal Methods 10(4):459–466

Zhang L, Qi Z, Zou Y, Zhang J, Xia W, Zhang R, Li J (2019) Engineering DNA–nanozyme interfaces for rapid detection of dental Bacteria. ACS Appl Mater Interfaces 11 (34):30640–30647

Zhao F, Wang Q, Dong J, Xian M, Yu J, Yin H, Wang J (2017) Enzyme-inorganic nanoflowers/ alginate microbeads: an enzyme immobilization system and its potential application. Process Biochem 57:87–94

Zheng Q, Wu H, Shen Z, Gao W, Yu Y, Ma Y, Ding K (2015) An electrochemical DNA sensor based on polyaniline/graphene: high sensitivity to DNA sequences in a wide range. Analyst 140 (19):6660–6670

Zhou J, Li H, Yang H, Cheng H, Lai G (2017) Immobilization of glucose oxidase on a carbon nanotubes/dendrimer-ferrocene modified electrode for reagent less glucose biosensing. J Nanosci Nanotech 17(1):212–216

Lectin Nanoconjugates for Targeted Therapeutic Applications

6

Mohini Verma, Ashish K. Shukla, and Amitabha Acharya

Contents

M. Verma · A. K. Shukla · A. Acharya (✉)
Biotechnology Division, CSIR-Institute of Himalayan Bioresource Technology, Palampur, Himachal Pradesh, India

Academy of Scientific & Innovative Research (AcSIR), CSIR-Institute of Himalayan Bioresource Technology, Palampur, Himachal Pradesh, India
e-mail: amitabha@ihbt.res.in

A. Acharya (ed.), *Nanomaterial - Based Biomedical Applications in Molecular Imaging, Diagnostics and Therapy*, https://doi.org/10.1007/978-981-15-4280-0_6

Abstract

Lectins are defined as a group of nonimmune proteins present mostly in plants (and also in some animals and microorganisms) that are known to bind with carbohydrates (mono- and oligosaccharides) with reversible and specific protein/animal cell agglutinating properties. These proteins are present in sufficient amounts to perform various crucial biological functions, viz., plants for defense, adhesion in bacteria, and in case of animals, binding of different proteins. Due to the efficiency and specificity of lectins to bind with carbohydrates, these are mostly used for various biomedical applications. Different types of carbohydrates get overexpressed in a number of diseases, ultimately acting as biomarkers for those diseases to be detected by the lectins. For getting better output, lectins are conjugated with materials that would provide a base for lectin's efficacy enhancement. After binding to a specific material, the randomness of the molecule gets decreased, which leads to an increase in the proficiency of the biomolecule. For this purpose, nanoparticles (NPs) act as the best platform for lectin conjugation. In general, the conjugation between the NPs and lectins takes place via covalent and non-covalent binding. All these linkages are preferably used in the binding strategies of these moieties. This binding of lectins to NPs helps to increase the specific targeting ability of the NPs by avoiding any nonspecific interaction. Due to higher specificity, NP-lectin conjugates find enormous applications in biomedical fields. In this chapter, we have tried to provide recent developments about these newly emerged nanoconjugates. A brief discussion is given on different lectin-NP conjugation techniques and its characterization followed by their biomedical applications.

Keywords

Carbohydrate binding · Lectins · Hemagglutination · Molecular imaging · Therapy

Abbreviations

AAL	*Aleuria aurantia* lectin
AGP	α1-Acid glycoprotein
ATU	3,9-Bis(3-aminopropyl)-2,4,8,10-tetraoxaspiro[5.5]undecane
AuNPs	Gold nanoparticles
BIST	Bead array in a single tip
CGB	Chempedak galactose-binding lectins
CNTs	Carbon nanotubes
Con A	Concanavalin A
CTAB	Cetyltrimethylammonium bromide
DSL	*Datura stramonium* Lectin
EDC	N-(3-Dimethylaminopropyl)-N-ethylcarbodiimide hydrochloride
ELISA	Enzyme-linked immunosorbent assay

EuNPs	Europium nanoparticles
MSN	Mesoporous silica nanoparticles
NHS	N-Hydroxysulfosuccinimide sodium salt
NPs	Nanoparticles
OL	Odorranalectin
PAA	Poly(acrylic acid)
PEG	Polyethylene glycol
PEG-MNPs	PEG-conjugated magnetic nanoparticles
PHA	Phytohemagglutinin
PLA	Polylactic acid
PLGA/PLG	Poly(lactic-co-glycolic acid)
PMBNR	Poly(MPC-co-n-butyl methacrylate (BMA)-co-o-nitrophenyloxycarbonyl polyethlene glycol methacrylate (MEONP)-co-p-methacryloxyethyl thiocarbonyl rhodamine B (MTR)
PNA	Peanut agglutinin
PVE	*Phaseolus vulgaris* erythroagglutinin
QDs	Quantum dots
TEOS	Tetraethyl orthosilicate
TEPSA	3-(Triethoxysilyl)propylsuccinic anhydride
TOPO	Trioctylphosphine
UE Lectins	*Ulex europaeus* lectin
WGA	Wheat germ agglutinin
ZnO NPs	Zinc oxide nanoparticles

6.1 Introduction

6.1.1 Brief of Nanotechnology

Nanotechnology introduces utilization as well as the creation of materials that exist at nanoscale ranging from 1 to 100 nm in size (Benelmekki 2019). It scrutinizes structural, magnetic, electrical, and optical behavior at molecular as well as submolecular levels. It has the capacity to revolutionize a series of biotechnology and medical tools to make them cheaper, compact, safer, and easily controllable. Nanoparticles (NPs) are tiny molecules with complex morphology consisting of a surface layer (with a wider range of metal ions, polymers, etc. for functionalization), followed by a shell layer, and finally the core, which is an essential part of the nanoparticle. Due to these characteristics, nanomaterials have acquired a huge interest in multidisciplinary work fields (Shin et al. 2016). Nowadays NPs play an important role in different fields to perform diverse functions like formation of oxide and solar fuel batteries for storage of energy, antimicrobial agent, sensors, and biological labeling agents in most of the disease treatments. Moreover, these are also used in daily requirements like clothes, cosmetics, etc. (Hasan 2015). NPs are known to be commonly synthesized by three different methods, i.e., physical, chemical, and biological process. Physical method includes attrition and pyrolysis for synthesizing nanoparticles, whereas the chemical method uses chemical techniques. Biosynthesis of nanomaterial is considered as an eco-friendly process (Mohanraj and Chen 2006). Advantage of biologically synthesized nanoparticles includes cost-effectiveness, lesser time, higher energy efficiency, and biocompatibility as compared to physical.

6.2 Outline of Lectins

Lectins are carbohydrate-binding proteins that are prevalently present in nature (Sharon and Lis 2004). Boyd and Shapleigh in 1954 coined the term lectin which means to choose/pick out. This name was given on the basis of its characteristic feature to distinguish different blood groups (Boyd and Shapleigh 1954; Hashim et al. 2017). This class of protein is seen in all biological systems playing a vital role like cell adhesion, cell agglutination, etc. Hemagglutination is a result of a reaction between lectins and carbohydrates present on the surface of erythrocytes (David et al. 2002). Ambrosi et al. described lectins as a tool for glycocode understanding (Ambrosi et al. 2005). Ricin, the first discovered lectin protein, was extracted from castor seeds (*Ricinus communis).* In 1919, James Summer purified crystalline lectin from jack bean *(Canavalia ensiformis*) with characteristic agglutination properties. Lectins play a major role in sugar transport/carbohydrate storage. It also contributes to symbiotic and pathogenic interaction between host and microbes. Lectin sensitivity results from its ability to cross certain protection barriers of the body, viz., secretory IgA (SIgA) that forms a barrier at the mucosal surface (Espino-Solis 2015). Based upon its molecular structure, lectins are classified into simple lectins, mosaic lectins, and macromolecular assembly. Simple lectins include lectins having a small number of subunits with a molecular weight below 40 KDa. All plant lectins that are found in legumes, cereals, Euphorbiaceae, and Moraceae and animal lectins including galectins and pentraxins come under this category. Mosaic lectins show diversity in structure and have different domains from which one module consists of the carbohydrate-binding site (e.g., viral hemagglutinins like influenza virus hemagglutinin, etc.). Macromolecular assembly includes bacterial lectins, viz., fimbriae and pili, which are filamentous and heteropolymeric organelles. The main part of the organelle is made up of polymer which contains a minor component to bind with carbohydrates. These lectins are also found in seeds, whereas some are also present in leaves, roots, flower, sap, bulb, and bark (Ting et al. 2010).

6.2.1 Classification of Lectins Based upon Source

6.2.1.1 Pant Lectins

These play a vital role in the plant defense system against fungal attack; moreover, it helps in the storage of proteins and modulation of enzymatic activity (Sharon 1993). These are used to determine the multivalent binding of carbohydrates. Interaction of carbohydrate and lectins requires the presence of Mn^{2+} and Ca^{2+} ions consisting of 2–4 subunits with typically 1 binding site per subunit. Concanavalin A (Con A (1919)), a model lectin, is used to examine the multivalent binding of glycopolymers. It is of low cost as it is found abundantly and easy to isolate and interacts with a wide range of saccharides (Pieters 2009). Con A is known to have binding affinity for mannose and glucose. Peanut agglutinin (PNA) has the ability to bind with galactosyl (β,1–3) N-acetylgalactosamine with high range of applications (Chacko and Appukutan 2001; Yilmaz and Becer 2015). It has an open quaternary structure, which is stabilized by hydrogen and hydrophobic interactions. Cereal lectins consist of two subunits, with two binding sites per unit. These are rich in disulfide bonds. Wheat germ agglutinin (WGA) has two subunits and is rich in cysteine (Tao et al. 2003). It exists in three different isoforms, WGA1, WGA2, and WGA3 having specificity toward N-acetylglucosamine and N-acetylneuraminic acid. Other lectins are found in bulb of plants like orchids, garlic, and amaryllis which bind to mannose. Jacalin, which is isolated from jackfruit of Moraceae family, can bind with galactose (β, 1–3) N-acetylgalactosamine and shows an interesting binding toward immunoglobulin A (Ertl et al. 2000). Another example is ricin, which can be isolated from beans of castor tree *(Ricinus communis)* and binds to galactose (Montisci et al. 2001). Ricin is a ribosomal inactivating protein that generates immunotoxins. Ricin and *Ricinus communis* agglutinin (RCA) are the most toxic lectins and cause immediate death (Ezpeleta et al. 1996b).

6.2.1.2 Animal Lectins

It comprises 12 different families, viz., C-type, I-type, P-type, S-type, trout lectins, calreticulin, ERGIC-53, VIP-36, fucolectins, pentraxins, discoidins, and fibrinogen-type lectins (Drickamer and Taylor 1993; Kilpatrick 2002). C-type lectins mostly include selectins, proteoglycans, and endocytic receptors. C-type lectins perform various functions, like that of cell-cell adhesion (Mitchell et al. 2001), immunogenic response, protein level control in blood, and also apoptosis. It may also be referred to as hepatic galactose/N-acetylglucosamine receptor (Lis and Sharon 1998; Sliedregt et al. 1999). Proteins can bind to different biomolecules, viz., proteins, lipids, and inorganic molecules, via lectins of C-type. C-type lectins may be found in the form of trimers, e.g., trimeric rat mannose-binding protein A in conjugation with α-methyl mannoside. It may exist in extremely small form, viz., pancreatic proteins of the intestine and also stone protein of the pancreas (Taylor et al. 2005). Selectins (C-type) act as cell adhesion molecules; the interaction of cells takes place via the coupling of selectins and glycol-conjugated ligands over the cell surface. Selectins are categorized into four subtypes, viz., E-selectins, present in endothelial cells; L-selectins, present in leukocytes; P-selectins, present in platelets and endothelial cells; and PSLG-1-selectins, present in hematopoietic cells. I-type lectin comprises immunoglobulin superfamily but excluding the receptors of T-cells as well as antibodies. Sialic acid-binding lectins are extracted from the Siglec lectin family (Brodaczewska et al. 2017). P-type lectins are known for the generation of functional lysosomes in higher eukaryotes as it helps to provide direction to the lysosomal enzymes which are freshly synthesized, bearing mannose 6-phosphate that gives an indication to lysosomes (Dahms and Hancock 2002). Pentraxin family comprises evolutionarily preserved pattern recognition proteins and is differentiated into short and long pentraxins. Short pentraxins include C-reactive proteins (CRP) and serum amyloid P-component (SAP). CRP has interaction potential with small ribonucleoproteins and can help in resolving infections and even autoimmune and cardiovascular diseases. SAP has the capability to identify nuclear substances and amyloid fibrils along with carbohydrates which later participate to resolve amyloidosis, etc. Galectins, discovered in 1976, are known to be isolated from chick muscle. It has tremendous activities starting from inflammation to cancer detection. Galectins show binding affinity with β-galactosides, especially for N-acetyllactosamine and lactose.

6.2.1.3 Microorganism Lectins

It includes lectins that are obtained from bacteria, viruses, fungi, etc. It was discovered in the 1950s by Alfred Gottschalk from the influenza virus (Nizet et al. 2017). Bacterial adhesion to cells depends on the molecular interactions that take place between the carbohydrates on the host cell surface and bacterial lectins. Adhesion by bacteria on the host cell takes place, so that it can withstand the defense mechanism of the host. Bacteria can form a colony on any biotic or abiotic surfaces due to the presence of fimbriae, hair-like lectins present on bacteria. The specificity of fimbriae depends over the type of bacteria that comprises different types, viz., type I lectins (fimbriae of type P, 17 kDa) are specific for Galα4Gal, and type II fimbriae present in oral actinomycetes show specificity for galactosides. *Pseudomonas aeruginosa* is known to form two types of lectin, viz., lecA (PA-IL) and lecB (PA-IIL) having specificity toward galactose and fucose, respectively (Soto and Hultgren 1999). Bacterial biofilm comprises a number of bacteria that are interconnected and communicate to each other via quorum sensing. Lectins help to adhere bacterial biofilm to the host cell, e.g., lectin (lecA and lecB) produced by *Pseudomonas aeruginosa* helps to bind with the host cell as these have affinity for the carbohydrate present on the biofilm matrix (da Silva et al. 2019). Bacteria also get protected from harsh environmental conditions due to the presence of biofilm.

6.3 Lectin-Nanoparticle Conjugation

Conjugation of lectins to different types of nanoparticles can be done by covalent attachment, non-covalent attachment, physical adsorption, or electrostatic interactions. Lectin nanoconjugates formed via covalent linking are more stable than formed via physical adsorption. Further, physical adsorption is beneficial for the proteins without the information of the specific target binding site. Commonly used conjugations include avidin-biotin binding (Tao et al. 2003); carbodiimide binding, in this carbodiimide catalyzes (Ertl et al. 2000); and glutaraldehyde coupling (Montisci et al. 2001; Ezpeleta et al. 1996a). Table 6.1 represents different strategies followed for lectin-NP conjugation and their respective applications.

Table 6.1 Different lectin-NP conjugation techniques and their application

Lectin	Nanoparticle	Conjugation	Application	References
PNA	PLGA	Glutaraldehyde	Oral immunization against hepatitis B	Gupta et al. (2006)
PHA	Chitosan	Biotin-avidin	Disease diagnosis	Liu et al. (2010)
Con A	AuNPs	EDC-NHS	Fungal detection	Olszewska et al. (2016)
WGA	PEG-MNPs	Streptavidin-biotin	*Staphylococcus aureus* detection	Yang et al. (2019)
Tomato lectin	Polystyrene (latex)	Carbodiimide	Contrast enhancement in MRI	Paschkunova-Martic et al. (2005)
WGA/PNA	PLGA	EDC-NHS	Treating inflammatory bowel disease	Moulari et al. (2014)
Cramoll lectin	QDs	Physical adsorption	Imaging of breast cancer	Carvalho et al. (2019)
WGA	QD-encapsulated PEG-PLA NPs	Thiol-melamide	Live cancer cell imaging	Gao et al. (2008b)
AAL	EuNPs	EDC-NHS	Prostate cancer detection	Kekki et al. (2017)
AAL, Con A	AuNPs	EDC-NHS	Detect glycosylation changes	Bose et al. (2018)
Con A	ZnONPs	EDC-NHS	Biosensing of arbovirusdetection	Simão et al. (2020)
WGA	QDs	Streptavidin-biotin	Cancer cell detection	Jeong et al. (2012)
WGA	Cyclodextrin-grafted liposomes	EDC-NHS	Treating oral cell	Wijetunge et al. (2020)
Cramoll lectin	QDs	Physical/covalent	Detect microbial infection	Cunha et al. (2018b)
WGA	QD-encapsulated PEG-PLA NPs	Thiol-maleimide	Brain imaging	Gao et al. (2008a)
WGA	CNTs	EDC-NHS	Biosensor for cancer cell	Yang et al. (2013)
WGA, macrophage galactose-type lectin	EuNPs	EDC-NHS	Cancer antigen 15–3 detection	Terävä et al. (2019)
UEA I/Con A	Gliadin nanoparticles	Carbodiimide	*Helicobacter pylori* treatment	Umamaheshwari and Jain (2003)
Sambucus nigra bark lectin	(PMBNR) + PLA	Avidin-biotin	Detection of sialic acid on cells	Cho et al. (2014)

6.3.1 Metallic Nanoparticle-Lectin Conjugation

6.3.1.1 Lectin-Gold Nanoparticle Conjugates

The lectin-conjugated AuNPs were formed by linking lectins to the exposed functional group on the surface of AuNPs through a bifunctional linker PEG NHS ester disulfide (NHS-PEG-S-S-PEG-NHS). Afterward, PEG-thiol (mPEG-SH) ligand was added in AuNP suspension to block targeted sites of AuNP which helps to enhance the stability of the lectin-AuNP conjugate. The reason behind using NHS-PEG-S-S-PEG-NHS was due to the internal S-S bonds of the compound that can be cleaved and chemically bonded onto the AuNP surface by Au-S formation. Again, the terminal group of the compound, viz., succinimidyl ester (NHS), enables it to react with primary and secondary amines on the lectin; and finally the PEG group behaves as a spacer and enhances activity that takes place after the interaction with the target (Wang et al. 2011; Hirsch et al. 2003; Loo et al. 2005; Lee et al. 2007).

6.3.1.2 Lectin-Conjugated Mesoporous Silica Nanoparticles

Mesoporous silica nanoparticles (MSN) were synthesized using the Stöber method in which tetraethyl orthosilicate (TEOS) acts as silica source and CTAB as a structure-directing agent. Functionalization of mesoporous silica NPs (MSN) with carboxylic group was done using 3-(triethoxysilyl)propylsuccinic anhydride (TEPSA) and acidified water, which helps to hydrolyze succinic groups to carboxylic groups. This MSN_{COOH} functionalization is followed by EDC-NHS coupling for grafting 3,9-bis(3-aminopropyl)-2,4,8,10-tetraoxaspiro[5.5]undecane (ATU) on its surface. Then this functionalized nanoparticle was modified using PAA which leads to the formation of MSN_{PAA}. Finally, Con A was covalently conjugated to MSN_{PAA} using carbodiimide chemistry in which amine group of protein forms bond with the carboxylic group of the nanoparticle (De Juan and Ruiz-Hitzky 2000; Schramm and Rinderer 2008; Sperling and Parak 2010; Mahon et al. 2012).

6.3.1.3 Lectin-Conjugated Quantum Dots

QDs were conjugated to WGA with the help of streptavidin-biotin conjugation (Jeong et al. 2012). In another study, QDs are conjugated to Cramoll lectin via physical adsorption (Cunha et al. 2018b).

6.3.2 Polymeric Nanoparticle-Lectin Conjugation

6.3.2.1 Lectin-Conjugated Vicilin Nanoparticles

Vicilin acts as a storage protein and is found in pea (*Pisum sativum*) and lentil. Nanoparticles were synthesized using this protein (Ezpeleta et al. 1996a) because of the presence of various functional groups for drug conjugation (Vermeersch and Remon 1994). Vicilin nanoparticles were conjugated to lectin by using covalent binding via two-stage carbodiimide method (Ezpeleta et al. 1996b; Damink et al. 1996).

6.3.2.2 Odorranalectin-Conjugated Nanoparticles

Odorranalectin (OL) is the smallest lectin and has less immunogenic response compared to other lectin family members. PEG-PLGA NPs were formed using double emulsification method, and these nanoparticles were mixed with OL to form OL-conjugated PEG-PLGA nanoparticles using maleimide (Mal-PEG-PLGA) reactions (Olivier et al. 2002; Wen et al. 2011; Gao et al. 2006).

6.3.2.3 Lectin-Conjugated Gliadin Nanoparticles

Gliadin, a cluster of polymorphic protein of gluten, is a plant storage protein found in wheat as well as in some cereals of grass genus. Gliadin protein is converted to gliadin nanoparticles (GNPs) via the desolvation process. EDC coupling was used before conjugating it with Con A/*Ulex europaeus* agglutinin I (Umamaheshwari and Jain 2003).

6.3.2.4 Lectin-Conjugated PEG-PLGA NPs

PEG-PLA NP-encapsulated QDs were synthesized, which were then conjugated to WGA for its biological applications. PEG-PLA nanoparticles were formed using methoxy-PEG-PLA and maleimide-PEG-PLA. Then, tri-n-octylphosphine oxide-coated CdSe/ZnS QDs (TOPO-QDs) were incorporated into the PEG-PLA nanoparticles via emulsion evaporation method. Later, WGA was conjugated with the QDs (Gao et al. 2008a; Yu et al. 2003; Zhang et al. 2004). This process is explained in Fig. 6.1.

6.4 Lectin Activity Studies Post-conjugation

6.4.1 Hemagglutination/Agglutination Assay

Hemagglutination assay is known to be the visual aggregation of cells and is the simplest way to get the semi-quantitative data for analyzing the conjugation of the lectin with the cell surface-bound carbohydrate of red blood cells (RBCs). The basic procedure for performing this assay is to dilute lectins serially in 96-well plates followed by the addition of erythrocytes in a constant volume. Erythrocytes that get interacted with lectins form a lattice that covers the well, whereas in other cases no such reaction takes place. The other way to perform this assay is to use a microscope glass slide directly, where RBCs and lectins were put together and seen under a microscope (Adamová et al. 2014). To improve the binding efficiency of lectin to glycoprotein, it can be conjugated to nanoparticles which help to enhance avidity and reduce false-positive results. Gliadin nanoparticles were conjugated to UEA or Con A lectin; this lectin nanoformulation showed agglutination when placed in *Helicobacter pylori* suspension. This proves the interaction of lectin-conjugated nanoparticle with bacterial cell wall containing mannose and α(L)-fucose residues (Umamaheshwari and Jain 2003). In another study WGA-conjugated PLGA NPs were also studied via hemagglutination assay (Yin et al. 2006).

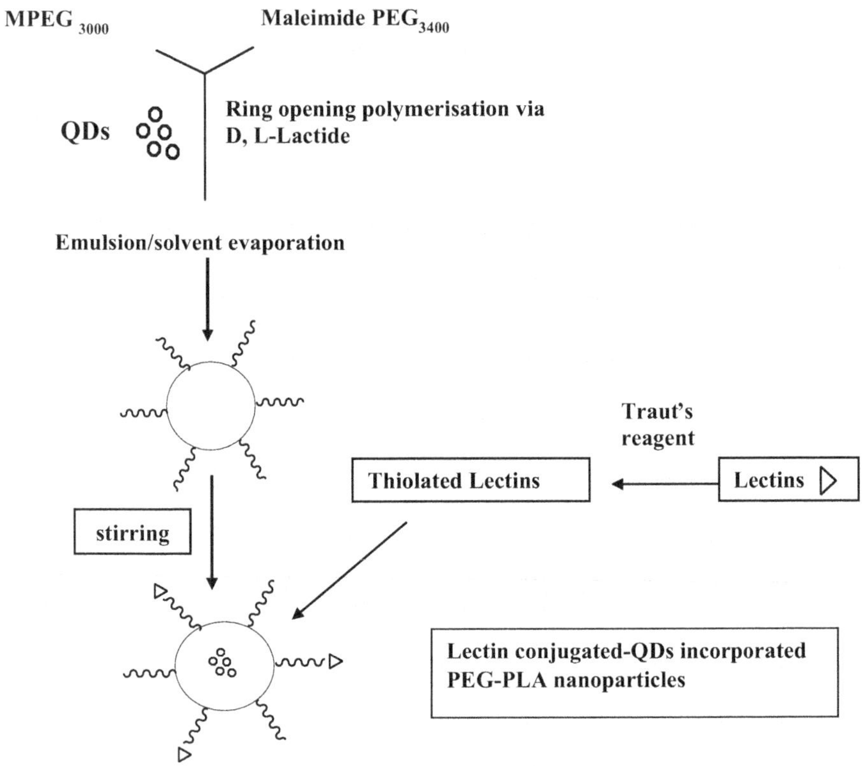

Fig. 6.1 Conjugation of lectins on quantum dot-encapsulated PEG-PLA nanoparticles

6.4.2 Carbohydrate-Lectin Inhibition Assay

This assay is used to check the activity of the conjugated lectin to the NPs. During the process of conjugation, there is possibility of inactivation of the lectin protein due to alterations in structure during the conjugation, attachment of the carbohydrate-binding site of lectin to the NPs surface, or non-conjugation of lectin. To confirm the activity of lectin after binding and conjugation accuracy, lectin-carbohydrate inhibition assay can be performed. After the formation of nanoconjugates, carbohydrate for which lectin shows specificity is added with the formed lectin nanoconjugate, and this undergoes hemagglutination assay (Bala Subramaniyan et al. 2019). Agglutination inhibition was seen by adding mannose and α(L)-fucose residues in the suspension of *Helicobacter pylori* before adding lectin nanoformulation (Con A and UEA lectin-conjugated gliadin nanoparticles) (Umamaheshwari and Jain 2003). Jeong et al. have performed inhibition study for WGA; in this N-acetylglucosamine (GlcNAc) was incubated with MCF-7 cells, and to this streptavidin-conjugated QDs were added (Jeong et al. 2012).

6.4.3 Fluorescence Quenching Studies

Fluorescence in proteins (lectins) originated due to the presence of aromatic amino acid residues, viz., tryptophan, tyrosine, and phenylalanine. During conjugation, there is a decrease in fluorescent intensity due to the Forster resonance energy transfer (FRET) mechanism (Bala Subramaniyan et al. 2019). Cunha et al., in their study, used QDs conjugated with Cramoll lectin and analyzed that there is peak shift after conjugation of lectin to nanoparticle (Cunha et al. 2018b).

6.5 Diagnostic and Therapeutic Applications

6.5.1 Enzyme-Linked Lectin Assay (ELLA)

An enzyme-linked lectin assay approach is a quantitative assay and is known to be a modified version of the ELISA technique and was used for almost three decades. The only difference between ELLA and ELISA is that the enzyme is linked to lectin instead of an antibody, in ELLA, for the identification of carbohydrates. This method comes in existence after the work done by McCoy (McCoy et al. 1983, 1985). ELLA has been done to identify oncofetal antigen galactosamine (glycoprotein) present in the sera of the patient having pancreatic cancer. Moreover, an enzyme-linked PNA assay was used to evaluate the level of Thomsen-Friedenreich antigen (T-Ag) in the serum of the patient having uterine cervix cancer. Quantification of the antigen in the patient's sera was done (Reddi et al. 2000; Ching and Rhodes 1989). Kekki et al. used ELLA with Eu-doped nanoparticles, and the assay was used in quantifying the presence of fucose over antigenic enzyme that is secreted in the prostate. This was tested using urine or blood sample. Comparative studies suggested that the concentration of Fucα1–6/3GlcNAc in prostate cancer is more than that found in benign prostatic hyperplasia (Kekki et al. 2017). Terävä et al. used the ELLA technique to detect cancer antigen 15–3 (CA15–3) in human plasma with the help of WGA-conjugated europium (Eu) NPs. It was proposed that lectin-conjugated NPs enhance multivalent binding to the glycan moiety of cancer antigen (Terävä et al. 2019).

6.5.2 Lectin Histochemistry

In lectin histochemistry, lectins are used in place of antibodies which are generally used in immunohistochemistry. Tissue staining involves the utilization of labeled lectins which help to detect only glycan-conjugated components. Lectin histochemistry provides information not only about the glycosylation processes occurring in tissues but also about the intracellular location. This is also helpful in disease diagnosis. Labeling in histochemistry can be done by both direct and indirect method. In direct labeling, lectins get labeled to fluorophores and different enzymes, whereas indirect method involves biotin/digoxigenin conjugation that can be diagnosed by enzyme-linked streptavidin or anti-digoxigenin (Roth 2011). Lectin histochemistry is used for studying glycosylation variations in cancerous tissues. There are two lectins (Con A and UEA-1) that are beneficial in differentiating the different histological grades of salivary gland cancer, i.e., mucoepidermoid carcinoma (Sobral et al. 2010). In parasite-induced cancer also known as cholangiocarcinoma, the biomarker, viz., glycoconjugate I paraffin was found in the patient's liver tissues, and this can be detected via multiple lectins (Indramanee et al. 2012). Andrade et al. evaluated the expression profile of α-D-glucose/mannose and L-fucose of normal, benign, and also malignant tissue of breast cancer by quantum dot lectin histochemistry. Two nanoformulations, viz., QD-conjugated Con A and QD-conjugated UEA I, were synthesized and used separately in this study. It has been proved that these nanoformulations have a distinct staining pattern which helps to distinguish the carbohydrate expression over normal, benign, and malignant breast cancer tissues (Andrade et al. 2013).

6.5.3 Lectin Blotting

This is known as the extension of Western blotting in which lectins are used instead of the antibody for glycoconjugate detection (Shan et al. 2001). Samples are resolved using polyacrylamide gel electrophoresis, followed by transfer over nitrocellulose membrane, and these were then detected via glycan-specific lectin probes. Previously this study was used to characterize the structure of glycan (Akama and Fukuda 2006) and quantification as well as detection of O-glycosylated proteins (Roth et al. 2012). Glycosylated proteins of bio-fluids act as cancer biomarkers when studied using lectin blotting (Li et al. 2014; Kitamura et al. 2003). This method helps to diagnose different types of cancer like endometrial cancer, cervical cancer (Abdul-Rahman et al. 2007), breast cancer, bone cancer, nasopharyngeal cancer, ovary cancer (Mu et al. 2012), and prostate cancer (Jayapalan et al. 2012; Jayapalan et al. 2013).

6.5.4 Lectin Array

It is a technique developed for quick as well as profound diagnosis of glycans. It uses multiple lectins immobilized on a solid support for the detection of different glycoproteins/glycolipids in a single sample (Hirabayashi et al. 2011; Hsu et al. 2006). This display of lectins in a single sample helps to study the different binding interactions at a time. Here the solid surface used is pre-treated glass surface, over which lectins are conjugated by covalent or physical adsorption (Kuno et al. 2005). Based on variances in the glycosylation of proteins isolated from serum, different lectin array strategies have been used (Šunderić et al. 2016). Shah et al. demonstrated that the lectin-glycoprotein interaction in serum sample can differentiate three groups, i.e., healthy person, Barrett's esophagus patients, and patients having esophageal adenocarcinoma, with the help of mass spectroscopy-associated

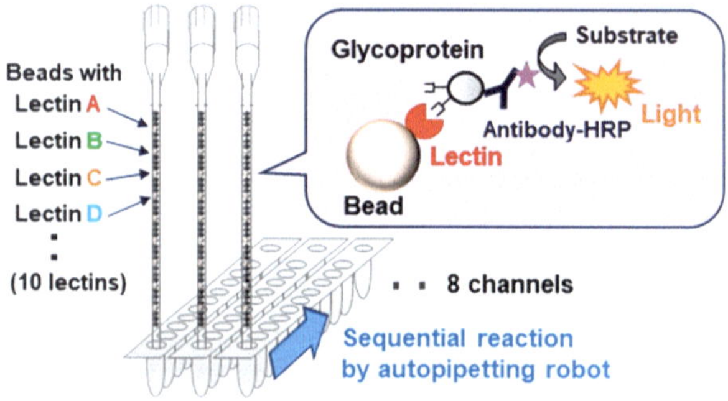

Fig. 6.2 Construction and concept of automatic glycoprotein analysis using the BIST system (GlycoBIST). (A) Illustration of the composition of a GlycoBIST tip and its position in the pipetting robot. (B) Schematic diagram of sequential reaction on a reaction bead. (C) Mixing protocol for the GlycoBIST reaction to produce an automatic batch interaction. (D) Image of scan data compared with the bead arrangement in the GlycoBIST tip (Adapted from Shimazaki et al. 2019, reprint with permission)

magnetic bead array (Shah et al. 2015). Lectin microarray analysis comes out as a prevailing method for analyzing disease-specific glycosylation variations in targeted proteins (Shimazaki et al. 2019). Gao et al. performed lectin assay for analyzing the specificity of four different bacterial strains, viz., *E. coli*, *E. cloacae*, *B. subtilis*, and *S. aureus*, and fungus, viz., *S. cerevisiae*, binding to 16 different types of lectins. Lectins were immobilized over glass slides and incubated with microbial strains, followed by addition of lectin-conjugated gold nanoparticles. This technique helps to distinguish different microbes based upon their structure (Gao et al. 2010b).

6.5.5 Lectin Bead Array in a Single Tip

Bead array in a single tip (BIST) is an automatic glycan profiling system. BIST is having a line ahead structural formation of 50 beads having 1 mm diameter in a capillary tip. As shown in Fig. 6.2, there are two types of beads, viz., spacer beads and reaction beads, which were separately formed and packed simultaneously line by line in a capillary tip. Lectins are immobilized over the reaction bead via physical adsorption. Multi-lectin assay by GlycoBIST was automatically performed by a Purelumn system robot (Kuno et al. 2011).

6.5.6 Bioimaging Applications

It has been reported that cancer causes alterations in normal cells and can be diagnosed via carbohydrate profiling of those cells/tissues. Analyzing the changes in glycobiology can become an effective tool in disease diagnosis. It is studied that lectins can bind specifically to their complementary carbohydrates and can be used to monitor and identify neoplastic changes in the body (Wormald and Sharon 2004; Oliveira et al. 2018). Literature reports suggest that lectins coupled to the fluorescent probe increase the sensitivity and specificity in in vivo as well as in vitro assays (Cunha et al. 2018a, b). Quantum dots can act as a fluorescent probe which can be used for conjugations (Schiffman and Balakrishna 2018). Quantum dots in conjugation with the lectin, viz., Cramoll (isolated from *Cratylia mollis* plant), have been reported to target and detect glucose and mannose glycocode in case of normal tissues, fibroadenoma, and invasive ductal carcinoma breast tissues. Binding specificity of lectins, when combined with the quantum dots, gives accurate information about the diseased site/condition. The use of a fluorescence microscope for accessing and evaluating the carbohydrate position was done. Detection and quantification of carbohydrate level were done using fluorescence microplate assay (FMA). From this study, it was concluded that the conjugates, viz., Cramoll QDs, act as an effective, specific probe that was able to differentiate between transformed cancerous and normal cells (Carvalho et al. 2019). Gao et al. developed WGA-conjugated quantum dots encapsulated in PEG-PLA NPs for tracing its specificity toward target cells. Cellular uptake of this lectin nanoformulation via clathrin- and caveolae-facilitated mechanisms was seen in Caco-2 cells, which takes place due to the binding of WGA to the cell surface carbohydrate moieties. It has been studied that this assembly showed fluorescence for almost 20 secs, whereas the dye-loaded NPs showed fluorescence for 0.5 s (Gao et al. 2008b). The same group worked with similar NPs for brain imaging, in which delivery of the nanoconjugates has been done through the nose. NPs formed were soluble in water, stable, and capable of brain targeting and imaging. The advantages of these nanoconjugates are as follows: first, it can stay in the brain for 4 h that can provide sufficient time to recognize as well as bind to the disorder region in the brain. Second, it can get cleared out after 8 h. Third, nanoconjugates get distributed rapidly in the brain as shown in Fig. 6.3 (Gao et al. 2008b). Further studies on lectin tagged polymeric nanoparticles (fluorescent) showed specificity for the sialic acid which is proven to be a biomarker in cancer cells, i.e., HeLa and MCF-7 (Cho et al. 2014). Lectin conjugation with Fe_2O_3@Au NPs was done, and this was used as a bioimaging probe for MR and CT imaging of the tumor. The conjugates formed show tremendous biocompatibility and stability in various media (including water, PBS, and culture media) and have better contrasting property. Moreover, in vivo studies proved that with the help of nanoconjugates there is an improvement in tumor MR imaging (He et al. 2014).

6.5.7 Biosensing Applications

Biosensors can be defined as the devices that are used for detecting the presence and quantification of biological analytes, like certain biomolecules, cellular entities, and also a plethora of microorganisms. A typical biosensor consists of a recognition element that recognizes the analyte and is able to produce the signal; a signal transduction element that generates the signal, with response corresponding to the analyte concentration; and a reader device that is able to read and monitor the signals

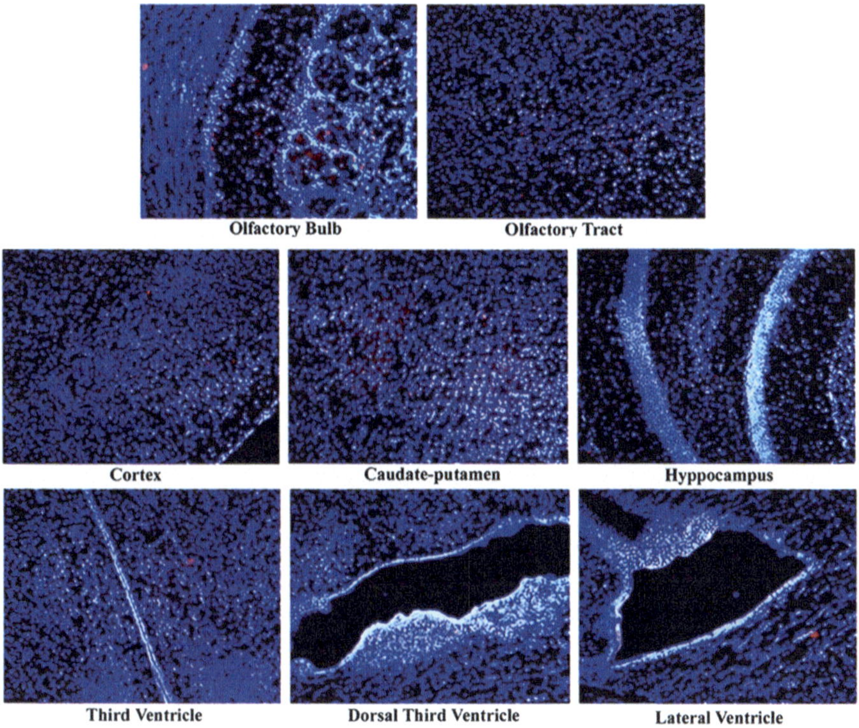

Fig. 6.3 Distribution of WGA-conjugated quantum dots-loaded nanoparticles (WGA-QD NP, red) in the olfactory bulb, olfactory tract, cortex, caudate-putamen, hippocampus, third ventricle, dorsal third ventricle, and lateral ventricle of nude mice 3 h after intranasal application (Adapted from Gao et al. 2008a, reprint with permission)

(Turner 2000). Biosensors can also be regarded as chemical sensors that are generally employed for biochemical detection of different classes of biomacromolecules, viz., antigens, proteins and peptides, enzymes, antibodies (Dar et al. 2019), and chemical analytes like pesticides (Dar et al. 2016; Walia and Acharya 2014). Based on the explorative properties of the NPs, miniaturized sensors that are inexpensive, highly selective, sensitive, and also able to measure the process in the static, as well as the continuous mode, are need of the moment. The NMs possess a number of unique properties that are highly sensitive and selective and may provide advanced and better opportunities in this field (Lim and Ramakrishna 2006). Lectins are carbohydrate-binding moieties that can be best suited to act as a biological receptor for increasing biosensor specificity. In the field of cancer, lectins are explored rapidly as a diagnostic tool. Nanomaterials, viz., metallic nanoparticles, as well as carbon nanotubes, are widely used in the formation of lectin-based biosensors as these helps to amplify the sensor signals. Con A-conjugated carbon nanotube electrochemical biosensor has been formed for the detection of glycans that are present over the live cancer cells. Passing through different comparative assays, it has been demonstrated that this biosensor was sensitive and selective toward the target and has a rapid analyzing response for mannose present on human lung cancer cell. It is suggested that this strategy can be used for other cancer studies using different lectins which are specific toward that cancer type (Zhang et al. 2011). Con A shows selectivity for glucose, making it an appropriate recognizing molecule for the development of glucose biosensor. Labib et al. developed a capacitive glucose-sensing biosensor with the conjugation of Con A on a gold nanoparticle-coated electrode having a glucose detection capacity of 1.0×10^{-6}–1.0×10^{-2} M (Labib et al. 2010). Con A-based magnetoelastic sensors were formed with the use of iron oxide nanoparticles which are modified with chitosan and showed significant sensing for *E. coli.* The number of the microbe attached can be analyzed by noticing the resonance frequency changes taking place in the probe (Lu et al. 2009; Lin et al. 2010).

6.5.8 Lectin-Mediated Bacterial Capture

Lectin-functionalized interfaces have been used for isolating bacteria from solution as this is a cost-effective and nondestructive method (Thepwiwatjit et al. 2014). Physical and chemical parameters were studied so that more and more bacteria can be captured (Tansub et al. 2012). Cell capture method has been used for pathogenic bacteria from food and environmental samples (Campuzan et al. 2012; Aden et al. 2017) and also for sepsis therapy in which microbes are removed from whole blood (Zhuang et al. 2017). For the formation of bacterial cell capture nanosensor, nanopillars on silicon surface were formed. To this, lectins conjugated azlactone-functional block copolymer $PGMA_{56}$ – b-$PVDMA_{\mathbf{175}}$ were coated (Masigol et al. 2019). On the basis of microarray principle, Gao et al. proposed a technique for bacterial capture. IgG antibody was immobilized over a glass slide which is functionalized using epoxide. The reason behind using monoclonal antibody is broad binding specificity due to which different bacteria can be targeted. Lectins, viz., *Ricinus communis* agglutinin (RCA) and concanavalin A (Con A), specifically bind to gram-negative and gram-positive bacteria, respectively. These lectins were conjugated to gold NPs followed by silver deposition (to enhance the signal). Detection is based upon the resonance light scattering (RLS) principle. This technique can help to discriminate between the strain types, viz., gram-negative and gram-positive, with a limit of 10^3 cells/assay. Selectivity for the strain can be managed by changing the lectins that are conjugated to gold NPs (Gao et al. 2010a).

6.5.9 Targeted Drug Delivery

The modern and advanced system of targeted delivery of drugs is a method of increasing the specificity of the drugs toward the target cells/tissues. This involves increasing the concentration of drugs in particular tissues or cells compared to the other non-targeted ones (Krishnan 2010). Nanomedicine has solved many issues with respect to drugs (Mishra et al. 2016). Due to unique properties (small size, increased surface area) of nanoparticles, hydrophobic drug delivery can be done by loading drugs to NPs. Other important and exclusive properties of nanomedicines are the stealth effect and liver metabolism bypass, and these require comparatively low dosage with respect to marketed drugs. Nanomedicines include liposomes, quantum dots (Misra 2008), dendrimers (Swami et al. 2012), etc. as drug delivery agents. Nano-sized molecules can be easily absorbed in comparison to microparticles (Sezer 2012). Nanomedicine has been reported for many diseases in which normal drugs are unable to reach, viz., inflammatory bowel disease (inflammation in the colon and small intestine). Lectin-conjugated nanoparticles, in which glucocorticoid was loaded, showed selectivity toward inflamed tissues in colitis. Here lectins (WGA, PNA) were bounded covalently. When a comparison was done between lectin-conjugated and plain NPs, the former showed better results especially PNA-conjugated NPs (Moulari et al. 2014). In other study, *Cratylia mollis* lectin (Cra), specific for mannose and glucose, was enclosed in liposome. Antitumor activity against sarcoma 180 was analyzed using these nanoformulations. It was seen that successful encapsulation (84%) of lectin (Cra) in liposomes retains its activity and enhanced protein stability as well as delivery. Moreover it shows 71% tumor inhibition. Encapsulated lectin shows less toxic effect on the kidney and liver compared to the bare one (Andrade et al. 2004). Paclitaxel- and isopropyl myristate-encapsulated PLGA nanoparticles, which were coated with WGA, were synthesized. These act as nanomedicine which helps to deliver paclitaxel to the lung as conjugation of WGA to nanoparticle makes it more capable to be taken up by cells easily. Its activity was checked against A549 and H1299 cells. It showed a much better result compared to the pure paclitaxel (Mo and Lim 2004; Mo and Lim 2005). The same kind of nanoparticles (paclitaxel-loaded WGA-coated PLGA) was formed for studying anticancer activity and stealth effect in colon cancer cells. WGA-conjugated NPs (WNP) showed an inhibitory effect against Caco-2 and HT-29 cells. This is possibly due to the presence of N-acetyl-D-glucosamine comprising carbohydrate-binding protein which gets overexpressed over the colon cell membrane (Park et al. 2012). Another benefit of this nanoparticle formulation is that it can bypass the endo-lysosome compartment that ultimately helps to increase its retention time inside the cell (Wang et al. 2010). More literature support was found in which nanoparticles act as therapeutic agents. Jacalin was conjugated to PEGylated gold nanoparticles that help to target Thomsen-Friedenreich disaccharide antigen (T-antigen) present on the surface of HT-29 cells of colorectal adenocarcinoma. As these NPs carry photosensitizer, viz., zinc phthalocyanine which helps in photodynamic therapy, 95–98% killing of cancer cells were observed in comparison to non-conjugated and nonirradiated NPs (Obaid et al. 2012). In another study, it has been reported that magnetic nanoparticles loaded with paclitaxel, after conjugation with lectin, gets engulfed through receptor-mediated endocytosis. This nanoformulation acts as the new theranostic approach for chronic myeloid leukemia (Singh et al. 2011). Yin et al. reported the delivery of thymopentin (immune stimulator) with the help of PLGA-NPs conjugated with WGA. The specificity of lectins was retained after conjugation. Moreover, WGA conjugation on NPs helps to increase the oral uptake of these nanoformulations by absorption of NPs and avoiding degradation of thymopentin (Yin et al. 2006). WGA-NP conjugate helps to examine the target delivery of β-galactosidase in the intestinal mucosa of rat. Binding of NPs to target site, viz., mucosa of the intestine, has been analyzed by a fluorometric method. It has been seen that lectin-conjugated NPs stayed for about 6.7 h which is 6.9-folds higher than the NPs without lectin. WGA-NPs may act as efficient conjugation for delivering mucosal drugs to treat lactose intolerance (Sheng et al. 2014).

6.6 Conclusion and Future Prospective

Development and progression of any disease lead to alteration on the surface of a cell which later acts as biomarkers of that disease. Once we detect that biomarker, it becomes somewhat possible to diagnose that disease so that it can be prevented from getting drastically uncontrollable. The specificity of a molecule is itself a unique phenomenon that helps to avoid false-positive results and also helps to screen out its complementary binding molecule. Binding between both these molecules plays a significant role. Different carbohydrate moieties get overexpressed in the disease condition which gets identified via the use of different types of lectin molecules extracted from either the plant, animal, or microorganism. Binding between lectin and carbohydrate acts as a lock and key hypothesis. Due to this property of efficiently binding, lectin is in existence and used as a diagnostic probe for diseases. For making the lectins more effective, these were conjugated to the nanoparticles, which help to provide a base for the lectins so that it can efficiently bind to the carbohydrates. Nanoparticles due to its more surface area allow more and more lectins to get conjugated and also show a multivalence effect due to which there is an enhancement in the interaction of lectins to carbohydrates. Different conjugation approaches have been discussed in detail which become a platform for the formation of different diagnostic techniques. Conjugation includes covalent and non-covalent binding. Covalent binding is much more preferable due to higher strength and specificity purpose. Covalent binding includes avidin-biotin technology, carbodiimide conjugation, and also glutaraldehyde binding. In the formation of lectin array, lectin bead array comes in existence due to specificity of lectins against carbohydrates overexpressed in cancer. In the future glycan profiling can be used for other diseases to find out the carbohydrate moiety that gets overexpressed in diseased condition. Bacteria expressing lectins can be engineered in such a way that they can act as attenuated vaccines in case of bacterial or biofilm infection. Many more applications may come into existence if get explored.

Acknowledgments The authors would like to thank the director of CSIR-IHBT for his constant support and encouragement. AA acknowledges financial assistance in the form of project grant MLP-0201 from CSIR and GAP-0214 (EMR/2016/003027) from DST, government of India. MV and AKS acknowledge DST and CSIR-GATE, respectively, for their fellowships. The CSIR-IHBT communication number of this manuscript is 4563.

References

Abdul-Rahman PS, Lim BK, Hashim OH (2007) Expression of high-abundance proteins in sera of patients with endometrial and cervical cancers: analysis using 2-DE with silver staining and lectin detection methods. Electrophoresis 28(12):1989–1996

Adamová L, Malinovská L, Wimmerová M (2014) New sensitive detection method for lectin hemagglutination using microscopy. Microsc Res Tech 77(10):841–849

Aden B, Kite CM, Hopkins BW, Zetterberg A, Lokitz BS, Ankner JF, Kilbey SM (2017) Assessing chemical transformation of reactive, interfacial thin films made of end-tethered poly (2-vinyl-4, 4-dimethyl azlactone) (PVDMA) chains. Macromolecules 50(2):618–630

Akama TO, Fukuda MN (2006) N-glycan structure analysis using Lectins and an α-Mannosidase activity assay. Methods Enzymol 416:304–314

Ambrosi M, Cameron NR, Davis BG (2005) Lectins: tools for the molecular understanding of the glycocode. Org Biomol Chem 3(9):1593–1608

Andrade CA, Correia MT, Coelho LC, Nascimento SC, Santos-Magalhães NS (2004) Antitumor activity of Cratylia mollis lectin encapsulated into liposomes. Int J Pharm 278(2):435–445

Andrade CG, Cabral Filho PE, Tenório DP, Santos BS, Beltrão EI, Fontes A, Carvalho LB Jr (2013) Evaluation of glycophenotype in breast cancer by quantum dot-lectin histochemistry. Int J Nanomedicine 8:4623–4629

Bala Subramaniyan S, Senthilnathan R, Arunachalam J, Anbazhagan V (2019) Revealing the significance of glycan binding property of butea monosperma seed lectin for enhancing the antibiofilm activity of silver nanoparticles against uropathogenic Escherichia coli. Bioconjug Chem 31:139–148

Benelmekki M (2019) Nanomaterials: the original product of nanotechnology. Morgan & Claypool Publishers, San Rafael

Bose PP, Chatterjee U, Chatterjee BP (2018) Assessment of altered Fucosylation of serum A-1-acid glycoprotein in hepatocellular carcinoma patients by gold-nanoparticle aggregation immunoassay. Mat Sci Res 15(3):263–271

Boyd WC, Shapleigh E (1954) Specific precipitating activity of plant agglutinins (lectins). Science 119(3091):419

Brodaczewska K, Wolaniuk N, Lewandowska K, Donskow-Łysoniewska K, Doligalska M (2017) Biodegradable chitosan decreases the immune response to Trichinella spiralis in mice. Molecules 22(11):2008

Campuzano S, Orozco J, Kagan D, Guix M, Gao W, Sattayasamitsathit S, Wang J (2012) Bacterial isolation by lectin-modified microengines. Nano Lett 12(1):396–401

Carvalho MET, Oliveira WF, Cunha CR, Coelho LC, Silva MV, Junior LBC, Correia MTS (2019) Evaluating the glycophenotype on breast cancer tissues with quantum dots-Cramoll lectin conjugates. Int J Biol Macromol 138:302–308

Chacko BK, Appukuttan PS (2001) Peanut (Arachis hypogaea) lectin recognizes α-linked galactose, but not N-acetyl lactosamine in N-linked oligosaccharide terminals. Int J Biol Macromol 28(5):365–371

Ching CK, Rhodes JM (1989) Enzyme-linked PNA lectin binding assay compared with CA19-9 and CEA radioimmunoassay as a diagnostic blood test for pancreatic cancer. Br J Cancer 59 (6):949

Cho J, Kushiro K, Teramura Y, Takai M (2014) Lectin-tagged fluorescent polymeric nanoparticles for targeting of sialic acid on living cells. Biomacromolecules 15(6):2012–2018

Cunha CRA, Oliveira ADPR, Firmino TVC, Tenório DPLA, Pereira G, Carvalho LB Jr, Fontes A (2018a) Biomedical applications of glyconanoparticles based on quantum dots. Biochim Biophys Acta 1862(3):427–439

Cunha CR, Andrade CG, Pereira MI, Cabral Filho PE, Carvalho LB Jr, Coelho LC, Correia MT (2018b) Quantum dot–Cramoll lectin as novel conjugates to glycobiology. J Photoch Photobio B 178:85–91 da Silva DP, Matwichuk ML, Townsend DO, Reichhardt C, Lamba D, Wozniak DJ, Parsek MR (2019) The Pseudomonas aeruginosa lectin LecB binds to the exopolysaccharide Psl and stabilizes the biofilm matrix. Nat Commun 10(1):1–11

Dahms NM, Hancock MK (2002) P-type lectins Biochim. Biophys Acta Gen Subj 1572 (2–3):317–340

Damink LO, Dijkstra PJ, Van Luyn MJA, Van Wachem PB, Nieuwenhuis P, Feijen J (1996) In vitro degradation of dermal sheep collagen cross-linked using a water-soluble carbodiimide. Biomaterials 17(7):679–684

Dar AI, Walia S, Acharya A (2016) Citric acid-coated gold nanoparticles for visual colorimetric recognition of pesticide dimethoate. J Nanopart Res 18(8):233

Dar AI, Walia S, Acharya A (2019) Molecular recognition based rapid diagnosis of immunoglobulins via proteomic profiling of protein-nanoparticle complexes. Int J Biol Macromol 138:156–167

David A, Kopecková P, Kopecek J, Rubinstein A (2002) The role of galactose, lactose, and galactose valency in the biorecognition of N-(2-hydroxypropyl) methacrylamide copolymers by human colon adenocarcinoma cells. Pharm Res 19(8):1114–1122

De Juan F, Ruiz-Hitzky E (2000) Selective functionalization of mesoporous silica. Adv Mat 12 (6):430–432

Drickamer K, Taylor ME (1993) Biology of animal lectins. Annu Rev Cell Biol 9(1):237–264

Ertl B, Heigl F, Wirth M, Gabor F (2000) Lectin-mediated bioadhesion: preparation, stability and caco-2 binding of wheat germ agglutinin-functionalized Poly (D, L-lactic-co-glycolic acid)-microspheres. J Drug Target 8(3):173–184

Espino-Solis GP (2015) Lectins: a brief review. Vitae 22(1):9–11

Ezpeleta I, Irache JM, Stainmesse S, Chabenat C, Gueguen J, Orecchioni AM (1996a) Preparation of lectin-vicilin nanoparticle conjugates using the carbodiimide coupling technique. Int J Pharm 142(2):227–233

Ezpeleta I, Irache JM, Stainmesse S, Gueguen J, Orecchioni AM (1996b) Preparation of small-sized particles from vicilin (vegetal protein from Pisum sativum L.) by coacervation. Eur J Pharm Biopharm 42(1):36–41

Gao X, Tao W, Lu W, Zhang Q, Zhang Y, Jiang XFS (2006) Lectin-conjugated PEG–PLA nanoparticles: preparation and brain delivery after intranasal administration. Biomaterials 27 (18):3482–3490

Gao X, Chen J, Chen J, Wu B, Chen H, Jiang X (2008a) Quantum dots bearing lectin-functionalized nanoparticles as a platform for in vivo brain imaging. Bioconjug Chem 19(11):2189–2195

Gao X, Wang T, Wu B, Chen J, Chen J, Yue Y, Jiang X (2008b) Quantum dots for tracking cellular transport of lectin-functionalized nanoparticles. Biochem Biophys Res Commun 377(1):35–40

Gao J, Liu C, Liu D, Wang Z, Dong S (2010a) Antibody microarray-based strategies for detection of bacteria by lectin-conjugated gold nanoparticle probes. Talanta 81(4–5):1816–1820

Gao J, Liu D, Wang Z (2010b) Screening lectin-binding specificity of bacterium by lectin microarray with gold nanoparticle probes. Anal Chem 82(22):9240–9247

Gupta PN, Mahor S, Rawat A, Khatri K, Goyal A, Vyas SP (2006) Lectin anchored stabilized biodegradable nanoparticles for oral immunization: 1 Development and in vitro evaluation. Int J Pharm 318(1–2):163–173

Hasan S (2015) A review on nanoparticles: their synthesis and types. Res J Rec Sci 4:1–3

Hashim OH, Jayapalan JJ, Lee CS (2017) Lectins: an effective tool for screening of potential cancer biomarkers. Peer J 5:e3784

He X, Liu F, Liu L, Duan T, Zhang H, Wang Z (2014) Lectin-conjugated Fe2O3@ Au core@ shell nanoparticles as dual mode contrast agents for in vivo detection of tumor. Mol Pharm 11 (3):738–745

Hirabayashi J, Kuno A, Tateno H (2011) Lectin-based structural glycomics: a practical approach to complex glycans. Electrophoresis 32(10):1118–1128

Hirsch LR, Jackson JB, Lee A, Halas NJ, West JL (2003) A whole blood immunoassay using gold nanoshells. Anal Chem 75(10):2377–2381

Hsu K-L, Pilobello KT, Mahal LK (2006) Analyzing the dynamic bacterial glycome with a lectin microarray approach. Nat Chem Biol 2(3):153–157

Indramanee S, Silsirivanit A, Pairojkul C, Wongkham C, Wongkham S (2012) Aberrant glycosylation in cholangiocarcinoma demonstrated by lectin-histochemistry. Asian Pac J Cancer Prev 13:119–124

Jayapalan JJ, Ng KL, Razack AH, Hashim OH (2012) Identification of potential complementary serum biomarkers to differentiate prostate cancer from benign prostatic hyperplasia using gel-and lectin-based proteomics analyses. Electrophoresis 33(12):1855–1862

Jayapalan JJ, Ng KL, Shuib AS, Razack AH, Hashim OH (2013) Urine of patients with early prostate cancer contains lower levels of light chain fragments of inter-alpha-trypsin inhibitor and saposin B but increased expression of an inter-alpha-trypsin inhibitor heavy chain 4 fragment. Electrophoresis 34(11):1663–1669

Jeong HH, KimYG JSC, Yi H, Lee CS (2012) Profiling surface glycans on live cells and tissues using quantum dot-lectin nanoconjugates. Lab Chip 12(18):3290–3295

Kekki H, Peltola M, van Vliet S, Bangma C, van Kooyk Y, Pettersson K (2017) Improved cancer specificity in PSA assay using Aleuria aurantia lectin coated Eu-nanoparticles for detection. Clin Biochem 50(1–2):54–61

Kilpatrick DC (2002) Animal lectins: a historical introduction and overview. Biochim Biophys Acta Gen Subj 1572(2–3):187–197

Kitamura N, Guo S, Sato T, Hiraizumi S, Taka J, Ikekita M, Sawada S, Fujisawa H, d Furukawa K (2003) Prognostic significance of reduced expression of β-N-acetylgalactosaminylated N-linked oligosaccharides in human breast cancer. Int J Cancer 105(4):533–541

Krishnan KM (2010) Biomedical nanomagnetics: a spin through possibilities in imaging, diagnostics, and therapy. IEEE Trans Magnetics 46(7):2523–2558

Kuno A, Uchiyama N, Koseki-Kuno S, Ebe Y, Takashima S, Yamada M, Hirabayashi J (2005) Evanescent-field fluorescence-assisted lectin microarray: a new strategy for glycan profiling. Nat Methods 2(11):851

Kuno A, Ikehara Y, Tanaka Y, Angata T, Unno S, Sogabe M, Narimatsu H (2011) Multilectin assay for detecting fibrosis-specific glyco-alteration by means of lectin microarray. Clin Chem 57 (1):48–56

Labib M, Hedström M, Amin M, Mattiasson B (2010) A novel competitive capacitive glucose biosensor based on concanavalin A-labeled nanogold colloids assembled on a polytyramine-modified gold electrode. Anal Chim Acta 659(1–2):194–200

Lee S, Kim S, Choo J, Shin SY, Lee YH, Choi HY, Oh CH (2007) Biological imaging of HEK293 cells expressing PLCγ1 using surface-enhanced Raman microscopy. Anal Chem 79(3):916–922

Li N, Dong G, Wang S, Zhu S, Shen Y, Li G (2014) Pinellia pedatisecta agglutinin-based lectin blot analysis distinguishes between glycosylation patterns in various cancer cell lines. Oncol Lett 8 (2):837–840

Lin H, Lu Q, Ge S, Cai Q, Grimes CA (2010) Detection of pathogen Escherichia coli O157: H7 with a wireless magnetoelastic-sensing device amplified by using chitosan-modified magnetic Fe3O4 nanoparticles. Sens Actuators B Chem 147(1):343–349

Lim TC, Ramakrishna S (2006) A conceptual review of nanosensors. Z Naturforsch A61(7–8):402–412

Lis H, Sharon N (1998) Lectins: carbohydrate-specific proteins that mediate cellular recognition. Chem Rev 98(2):637–674

Liu J, Zhang L, Wang C, Xu H, Zhao X (2010) Preparation and characterization of lectin-conjugated chitosan fluorescent nanoparticles. Mol Biosyst 6(6):954–957

Loo C, Lowery A, Halas N, West J, Drezek R (2005) Immunotargeted nanoshells for integrated cancer imaging and therapy. Nano Lett 5(4):709–711

Lu Q, Lin H, Ge S, Luo S, Cai Q, Grimes CA (2009) Wireless, remote-query, and high sensitivity Escherichia coli O157: H7 biosensor based on the recognition action of concanavalin A. Anal Chem 81(14):5846–5850

Mahon E, Salvati A, Bombelli FB, Lynch I, Dawson KA (2012) Designing the nanoparticle–biomolecule interface for "targeting and therapeutic delivery". J Control Release 161 (2):164–174

Masigol M, Fattahi N, Barua N, Lokitz BS, Retterer ST, Platt TG, Hansen RR (2019) Identification of critical surface parameters influencing lectin-mediated capture of Bacteria from solution. Biomacromolecules 20:2852–2863

McCoy JP Jr, Varani J, Goldstein IJ (1983) Enzyme-linked lectin assay (ELLA): use of alkaline phosphatase-conjugated Griffonia simplicifolia B4 isolectin for the detection of α-D-galactopyranosyl end groups. Anal Biochem 130(2):437–444

McCoy JJ, Goldstein IJ, Varani J (1985) A review of studies in our laboratory regarding ELLA methodology for the study of cell surface carbohydrates from tumors of varying metastatic potential. Tumour Biol 6(2):99–114

Mishra N, Pant P, Porwal A, Jaiswal J, Aquib M (2016) Targeted drug delivery: a review. Am J Pharm Tech Res 6:2249–3387

Misra RD (2008) Quantum dots for tumor-targeted drug delivery and cell imaging. Nanomedicine 3 (3):271–274

Mitchell DA, Fadden AJ, Drickamer K (2001) A novel mechanism of carbohydrate recognition by the C-type lectins DC-SIGN and DC-SIGNR subunit organization and binding to multivalent ligands. J Biol Chem 276(31):28939–28945

Mo Y, Lim LY (2004) Mechanistic study of the uptake of wheat germ agglutinin-conjugated PLGA nanoparticles by A549 cells. J Pharm Sci 93:20–28

Mo Y, Lim LY (2005) Preparation and in vitro anticancer activity of wheat germ agglutinin (WGA)-conjugated PLGA nanoparticles loaded with paclitaxel and isopropyl myristate. J Control Release 107:30–42

Mohanraj VJ, Chen Y (2006) Nanoparticles-a review. Trop J Pharm Res 5(1):561–573

Montisci MJ, Giovannuci G, Duchene D, Ponchel G (2001) Covalent coupling of asparagus pea and tomato lectins to poly (lactide) microspheres. Int J Pharm 215(1–2):153–161

Moulari B, Béduneau A, Pellequer Y, Lamprecht A (2014) Lectin-decorated nanoparticles enhance binding to the inflamed tissue in experimental colitis. J Control Release 188:9–17

Mu AKW, Lim BK, Hashim OH, Shuib AS (2012) Detection of differential levels of proteins in the urine of patients with endometrial cancer: analysis using two-dimensional gel electrophoresis and O-glycan binding lectin. Int J Mol Sci 13(8):9489–9501

Nizet V, Varki A, Aebi M (2017) Microbial lectins: hemagglutinins, adhesins, and toxins. In: Essentials of Glycobiology 3rd edition. Cold Spring Harbor Laboratory Press, Cold Spring Harbor

Obaid G, Chambrier I, Cook MJ, Russell DA (2012) Targeting the oncofetal Thomsen–Friedenreich disaccharide using Jacalin-PEG phthalocyanine gold nanoparticles for photodynamic cancer therapy. Angew Chem Int Ed 51(25):6158–6162

Oliveira WF, Silva GM, Cabral Filho PE, Fontes A, Oliveira MD, Andrade CA, Correia MT (2018) Titanium dioxide nanotubes functionalized with Cratylia mollis seed lectin, Cramoll, enhanced osteoblast-like cells adhesion and proliferation. Mater Sci Eng C 90:664–672

Olivier JC, Huertas R, Lee HJ, Calon F, Pardridge WM (2002) Synthesis of pegylated immunonanoparticles. Pharm Res 19(8):1137–1143

Olszewska A, Armstrong E, Olszewski M, Batenjany M, Ostafin A (2016) Lectin-based SERS sandwich immunoassay. J Nanomed Nanotechnol 7(405):2

Park SY, Lee SH, Kawasaki N, Itoh S, Kang K, Hee Ryu S, Hashii N, Kim JM, Kim JY, Hoe Kim J (2012) α1-3/4 fucosylation at Asn 241 of β-haptoglobin is a novel marker for colon cancer: a combinatorial approach for development of glycan biomarkers. Int J Cancer 130 (10):2366–2376

Paschkunova-Martic I, Kremser C, Mistlberger K, Shcherbakova N, Dietrich H, Talasz H, Zou Y, Hugl B, Galanski M, Sölder E, Pfaller K, Höliner I, Buchberger W, Keppler B, Debbage P (2005) Design, synthesis, physical and chemical characterisation, and biological interactions of

lectin-targeted latex nanoparticles bearing Gd–DTPA chelates: an exploration of magnetic resonance molecular imaging (MRMI). Histochem Cell Biol 123(3):283–301
Pieters RJ (2009) Maximising multivalency effects in protein–carbohydrate interactions. Org Biomol Chem 7(10):2013–2025
Reddi AL, Sankaranarayanan K, Arulraj HS, Devaraj N, Devaraj H (2000) Enzyme-linked PNA lectin-binding assay of serum T-antigen in patients with SCC of the uterine cervix. Cancer Lett 149(1–2):207–211
Roth J (2011) Lectins for histochemical demonstration of glycans. Histochem Cell Biol 136(2):117
Roth Z, Yehezkel G, Khalaila I (2012) Identification and quantification of protein glycosylation. Int J Carbohydr Chem 2012:1–10
Schiffman JD, Balakrishna RG (2018) Quantum dots as fluorescent probes: synthesis, surface chemistry, energy transfer mechanisms, and applications. Sensor Actuat B Chem 258:1191–1214
Schramm C, Rinderer B (2008) Investigation of the hydrolysis of (3-triethoxysilylpropyl) succinic acid anhydride by means of FT-IR. J Mat Sci 43(12):4215–4219
Sezer AD (ed) (2012) Recent advances in novel drug carrier systems. BoD–Books on Demand, 2012 Oct 31, ISBN: 978–953–51-4268-3
Shah AK, Lê Cao KA, Choi E, Chen D, Gautier B, Nancarrow D, Hill MM (2015) Serum glycoprotein biomarker discovery and qualification pipeline reveals novel diagnostic biomarker candidates for esophageal adenocarcinoma. Mol Cell Proteomics 14(11):3023–3039
Shan S, Tanaka H, Shoyama Y (2001) Enzyme-linked immunosorbent assay for glycyrrhizin using anti-glycyrrhizin monoclonal antibody and an eastern blotting technique for glucuronides of glycyrrhetic acid. Anal Chem 73(24):5784–5790
Sharon N (1993) Lectin-carbohydrate complexes of plants and animals: an atomic view. Trends Biochem Sci 18(6):221–226
Sharon N, Lis H (2004) History of lectins: from hemagglutinins to biological recognition molecules. Glycobiology 14(11):53R–62R
Sheng Y, He H, Zou H (2014) Poly(lactic acid) nanoparticles coated with combined WGA and water-soluble chitosan for mucosal delivery of β-galactosidase. Drug Deliv 21(5):370–378
Shimazaki H, Saito K, Matsuda A, Sawakami K, Kariya M, Segawa O, Kaji H (2019) Lectin bead array in a single tip facilitates fully automatic glycoprotein profiling. Anal Chem 91 (17):11162–11169
Shin WK, Cho J, Kannan AG, Lee YS, Kim DW (2016) Cross-linked composite gel polymer electrolyte using mesoporous methacrylate-functionalized SiO_2 nanoparticles for lithium-ion polymer batteries. Sci Rep 6:26332–26341
Simão EP, Silva DB, Cordeiro MT, Gil LH, Andrade CA, Oliveira MD (2020) Nanostructured impedimetric lectin-based biosensor for arboviruses detection. Talanta 208:120338
Singh A, Dilnawaz F, Sahoo SK (2011) Long circulating lectin conjugated paclitaxel loaded magnetic nanoparticles: a new theranostic avenue for leukemia therapy. PLoS One 6(11): e26803
Sliedregt LA, Rensen PC, Rump ET, van Santbrink PJ, Bijsterbosch MK, Valentijn ARP, Biessen EA (1999) Design and synthesis of novel amphiphilic dendritic galactosides for selective targeting of liposomes to the hepatic asialoglycoprotein receptor. J Med Chem 42(4):609–618
Sobral APV, Rego MJ, Cavalacanti CL, Carvalho LB Jr, Beltrão EI (2010) ConA and UEA-I lectin histochemistry of parotid gland mucoepidermoid carcinoma. J Oral Sci 52(1):49–54
Soto GE, Hultgren SJ (1999) Bacterial adhesins: common themes and variations in architecture and assembly. J Bacteriol 181(4):1059–1071
Sperling RA, Parak WJ (2010) Surface modification, functionalization and bioconjugation of colloidal inorganic nanoparticles. Philos Trans A Maths Phys Eng Sci 368(1915):1333–1383
Šunderić M, Šedivá A, Robajac D, Miljuš G, Gemeiner P, Nedić O, Katrlík J (2016) Lectin-based protein microarray analysis of differences in serum alpha-2-macroglobulin glycosylation between patients with colorectal cancer and persons without cancer. Biotechnol Appl Biochem 63(4):457–464

Swami A, Shi J, Gadde S, Votruba AR, Kolishetti N, Farokhzad OC (2012) Nanoparticles for targeted and temporally controlled drug delivery, Multifunctional nanoparticles for drug delivery applications. Springer, Boston, pp 9–29

Tansub W, Tuitemwong K, Limsuwan P, Theparoonrat S, Tuitemwong P (2012) Synthesis of antibodies-conjugated fluorescent dye-doped silica nanoparticles for a rapid single step detection of Campylobacter jejuni in live poultry. J Nanomater 2012:60

Tao SL, Lubeley MW, Desai TA (2003) Bioadhesive poly (methyl methacrylate) micro devices for controlled drug delivery. J Cont Rel 88(2):215–228

Taylor PR, Martinez-Pomares L, Stacey M, Lin HH, Brown GD, Gordon S (2005) Macrophage receptors and immune recognition. Annu Rev Immunol 23:901–944

Terävä J, Tiainen L, Lamminmäki U, Kellokumpu-Lehtinen PL, Pettersson K, Gidwani K (2019) Lectin nanoparticle assays for detecting breast cancer-associated glycovariants of cancer antigen 15-3 (CA15-3) in human plasma. PLoS One 14(7):1–16

Thepwiwatjit N, Thattiyaphong A, Limsuwan P, Tuitemwong K, Tuitemwong P (2014) Rubpy dye-doped silica nanoparticles as signal reporter in a dot fluorescence immunoassay strip. J Nanomater 2014:23

Ting SS, Chen G, Stenzel MH (2010) Synthesis of glycopolymers and their multivalent recognitions with lectins. Pol Chem 1(9):1392–1412

Turner AP (2000) Biosensors–sense and sensitivity. Science 290(5495):1315–1317

Umamaheshwari RB, Jain NK (2003) Receptor mediated targeting of lectin conjugated gliadin nanoparticles in the treatment of Helicobacter pylori. J Drug Target 11(7):415–424

Vermeersch H, Remon JP (1994) Immunogenicity of poly-D-lysine, a potential polymeric drug carrier. J Control Release 32(3):225–229

Walia S, Acharya A (2014) Fluorescent cadmium sulfide nanoparticles for selective and sensitive detection of toxic pesticides in aqueous medium. J Nanoparticle Res 16(12):2778

Wang C, Ho PC, Lim LY (2010) Wheat germ agglutinin-conjugated PLGA nanoparticles for enhanced intracellular delivery of paclitaxel to colon cancer cells. Int J Pharm 400(1–2):201–210

Wang J, Liu D, Wang Z (2011) Synthesis and cell-surface binding of lectin-gold nanoparticle conjugates. Anal Methods 3(8):1745–1751

Wen Z, Yan Z, Hu K, Pang Z, Cheng X, Guo L, Zhang Q, Jiang X, Fang L, Lai R (2011) Odorranalectin-conjugated nanoparticles: preparation, brain delivery and pharmacodynamic study on Parkinson's disease following intranasal administration. J Control Release 151 (2):131–138

Wijetunge SS, Wen J, Yeh CK, Sun Y (2020) Wheat germ agglutinin liposomes with surface grafted cyclodextrins as bioadhesive dual-drug delivery nanocarriers to treat oral cells. Colloids Surf B Biointerfaces 185:110572

Wormald MR, Sharon N (2004) Carbohydrates and glycoconjugates: progress in non-mammalian glycosylation, glycosyltransferases, invertebrate lectins and carbohydrate–carbohydrate interactions. Curr Opin Struct Biol 14(5):591–592

Yang H, Li Z, Wei X, Huang R, Qi H, Gao Q, Zhang C (2013) Detection and discrimination of alpha-fetoprotein with a label-free electrochemical impedance spectroscopy biosensor array based on lectin functionalized carbon nanotubes. Talanta 111:62–68

Yang G, Huang M, Wang Y, Chen G, Zhao Y, Xu H (2019) Streptavidin-exposed magnetic nanoparticles for lectin magnetic separation (LMS) of Staphylococcus aureus prior to three quantification strategies. Microchim Acta 186(12):813

Yilmaz G, Becer CR (2015) Glyconanoparticles and their interactions with lectins. Polym Chem 6 (31):5503–5514

Yin Y, Chen D, Qiao M, Lu Z, Hu H (2006) Preparation and evaluation of lectin-conjugated PLGA nanoparticles for oral delivery of thymopentin. J Control Release 116(3):337–345

Yu WW, Qu L, Guo W, Peng X (2003) Experimental determination of the extinction coefficient of CdTe, CdSe, and CdS nanocrystals. Chem Mat 15(14):2854–2860

Zhang Y, Zhang Q, Zha L, Yang W, Wang C, Jiang XFS (2004) Preparation, characterization and application of pyrene-loaded methoxy poly (ethylene glycol)–poly (lactic acid) copolymer nanoparticles. Colloid Polym Sci 282(12):1323–1328

Zhang X, Teng Y, Fu Y, Zhang S, Wang T, Wang C, Zhang W (2011) Lectin-based electrochemical biosensor constructed by functionalized carbon nanotubes for the competitive assay of glycan expression on living cancer cells. Chem Sci 2(12):2353–2360

Zhuang MY, Wang C, Xu MQ, Ling XM, Shen JJ, Zhang YW (2017) Using concanavalin A as a spacer for immobilization of E. coli onto magnetic nanoparticles. Int. J Biol Macromol 104:63–69

Plant-Based Polymeric Nanomaterials for Biomedical Applications

7

Syed M. S. Abidi, Aqib Iqbal Dar, and Amitabha Acharya

Contents

Abstract

Plant-derived polymeric nanomaterials play an important role as excipients in different biomedical applications ranging from nanomedicine to tissue engineering. These provide versatile biocompatible platforms for controlled drug release with the advantage of being most cost-effective materials. Inorganic nanoparticles

S. M. S. Abidi · A. I. Dar · A. Acharya (✉)
Biotechnology Division, CSIR-Institute of Himalayan Bioresource Technology, Palampur, H.P, India

Academy of Scientific & Innovative Research (AcSIR), CSIR-Institute of Himalayan Bioresource Technology, Palampur, H.P, India
e-mail: amitabha@ihbt.res.in

A. Acharya (ed.), *Nanomaterial - Based Biomedical Applications in Molecular Imaging, Diagnostics and Therapy*, https://doi.org/10.1007/978-981-15-4280-0_7

have received considerable global interest amongst biomedical researchers, but their toxicity concerns cannot be ignored. In this context, the biomass has been recognized as potential source to obtain different types of functional nanomaterials. These nanomaterials have polysaccharide backbone, which consists of monomer units of carbohydrate and mimic the extracellular matrix of tissues, thus making these biocompatible by nature. Further, the ease of surface functionalization, natural abundance, together with high stability make these ideal biomaterial candidates for next generation health care applications. Here, in this chapter, we have described the biomedical applications of different plant-based nanomaterials, with a major focus towards using these as drug delivery agents, bioprinting materials, wound healing agents, etc.

Keywords

Natural polymers · Polysaccharides · Bioprinting · Drug delivery agents · Tissue engineering

Abbreviations

[Bmim]PF6	1-butyl-3-methylimidazolium hexafluorophosphate
AcGGM	O-acetyl-galactoglucomannan
ATP	4-aminothiophenol
CMC	Carboxymethylcellulose
CS	Chitosan
DSS	Dentin phospho-phoryn-derived peptide
ECM	Extracellular matrix
FDA	Food and Drug Administration
GRA	Generally recognized as safe
HA	Hyaluronic acid
HACC	Hydroxypropyltrimethyl ammonium chloride chitosan
HEC	Hydroxyethyl cellulose
LLO91–99	Listeria monocytogenes Listeriolysin
LNP	Lignin nanoparticles
n- HA	Nano-hydroxyapatite
PBN	Plant-based nanomaterials
PCL	Polycaprolactone
PEG	Polyethlene glycol
PHB	Polyhydroxybutyrate
PLLA	Poly(l-lactide)
PVA	Polyvinyl alcohol
RCSP	Rana chensinensis skin peptides
SPI	Soy protein isolate
SPIONS	Superparamagnetic iron oxide nanoparticles
TEBAC	Triethylbenzylammonium chloride
TEMPO	2,2,6,6-Tetramethylpiperidine-1-oxyl radical

7.1 Introduction

Plant-based nanomaterials offer a new route for developing organic-based nanomaterials. These nanomaterials are of great interest among researchers because of their lower toxicity and enhanced physiochemical properties. The main objective behind using plant-based nanosytem is to enhance stability and efficiency, which can be achieved by converting them into the composite. The ease of surface functionalization, due to good chemical functionalities, tuned these nanomaterials as active excipients for different biomedical applications (Singla et al. 2017, 2019a, b; Xue et al. 2017). There are many literature reports which have shown surface functionalized nanomaterials exhibited better results under in vivo condition and played the crucial role in various biological application. These nanomaterials can be used in various forms, from nanocomposite-based platform of aerosol to hydrogels, which possess exclusive properties. This chapter covers the utility of different plant-based nanomaterial, i.e. cellulose, hemicellulose, lignin, starch, inulin, rosin, pectin and gum, etc. These parts also cover discussion of how these plant-based nanomaterials have been used for different animal studies. It is very difficult to document all the applications of plant-based nanomaterials in one chapter, but the present chapter covers a résumé of all these aspects.

7.2 Why Plant-Based Nanomaterials Are Preferred Over Conventional Synthetic and Inorganic Nanomaterials

Organic nanomaterials are engineered with multi-functionalities to pave the path for modern medicines and health care applications. Plant-derived nanomaterials offer biocompatible, stimuli-responsive, biodegradable and effective systems for targeted delivery of medicines. Advanced technology of top down (i.e. polymer to dispersion, viz. solvent evaporation, nanoprecipitation, salting out, dialysis) and bottom up (i.e. monomer to polymerization, viz. emulsion, surfactant free emulsion, nano-emulsion and interfacial polymerization) approaches have been utilized for the synthesis of size-controlled, multimodal, polymeric nanomaterials. Design of nanoparticle should be according to its biomedical applications and should be considered with different factors, like nature of materials and its mechanism of targeting towards disease site. The immense interest in polysaccharide-based bionanotechnological products requires detailed characterization and advanced synthesis methodologies, e.g. microfluids system, microelectron mechanical system, microneedle-based system, advanced high-pressure homogenization, electro piping system, etc. to impart specific physicochemical properties to the nanoparticles for exact application.

Plant based nanomaterial (PBN), systematically investigated worldwide, as biomaterials for regenerative medicines, because of their biochemical similarity with human extracellular matrix component, make these to be easily accepted by the body. The building blocks of these moieties consist of carbohydrate monosaccharides, which are joined together by O-glyosidic linkage, to form polysaccharide chains. These structural backbones provide diverse sets of chemical versatilities for various biomedical applications. The structure offers free hydroxyl and amine group, which make it possible to alter its physiochemical properties by chemically modifying polysaccharide structure (Yin et al. 2019). Moreover, reversion of synthetic to natural plant-based nanomaterials is required because these are easily available and produced by many countries. These are economically cheaper and fabrication cost is less than synthetic materials. PBNs are nontoxic and biocompatible, safe and devoid of any side effect, and its biodegradable nature makes it environment friendly. The new insights in this area could generate commercial health care technology and product from renewable resources and will contribute immensely towards economic development.

7.3 Sources of Plant-Based Nanomaterials

Diverse plant-derived nanomaterials confer different physiochemical properties. This variability and versatility was not present in any other class of nanomaterials and its ease of functionalization made it useful for advanced therapeutic applications. Impact of research has increased when the interest was shifted from inorganic to organic plant-based nanomaterials because there is a void space in the in-depth characterization of these materials. There are different types of nanomaterials that can be obtained from plants, viz., (a) nanocellulose or cellulose nanocrystals, which are a class of nanomaterial that are most abundant (7.5×10^{10} tons) in nature (Fig. 7.1). These materials are derived from flora that had renewable potential, which reduces environmental and ecological concerns. These nanomaterials are composed of β (1–4) linked D-glucose, which exhibit impressive physiochemical properties; (b) lignin, a second most abundant heterogenic complex phenolic biopolymer present on earth with potential antimicrobial properties; (c) hemicellulose, a non-crystalline heteropolymer, and its content varies among plant species. It is composed of 1,4 linked β-D-pyranol residues such as glucose, xylose (which is predominant) and mannose residues, in which oxygen is at fourth position in equilateral orientation; (d) starch, a polymeric carbohydrate composed of amylose and amylopectin glucose units joined by α 1,4 linkages. These are highly renewable and biodegradable polymer and its multiscale semi-crystalline structure provide two distinct nanomaterials, i.e. starch nanoparticles and starch nanocrystals; (e) pectin, a heteropolysaccharide, is composed of 65% homogalacturonan α-(1–4)-linked, 20–35% rhamnose, a small amount of arabinogalactan and xylogalacturonan. It has excellent bioavailability and bioactivity; (f) inulin, a Food and Drug Administration (FDA) approved plant polysaccharide. It is a linear polysaccharide composed of fructose units linked by glyosidic bonds and terminated at a single glucopyranoside ring because of which surface functionalization is very easy; moreover, it has excellent biocompatibility and biodegradability; (g) rosin, a natural biopolymer obtained from pine tree and composed of 90% rosin acids and 10% non-acidic materials and thus can be used for the preparation of different kinds of nanomaterials; (h) gums-based nanomaterials composed of polysaccharide backbone with different physiochemical and mechanical properties. There are different

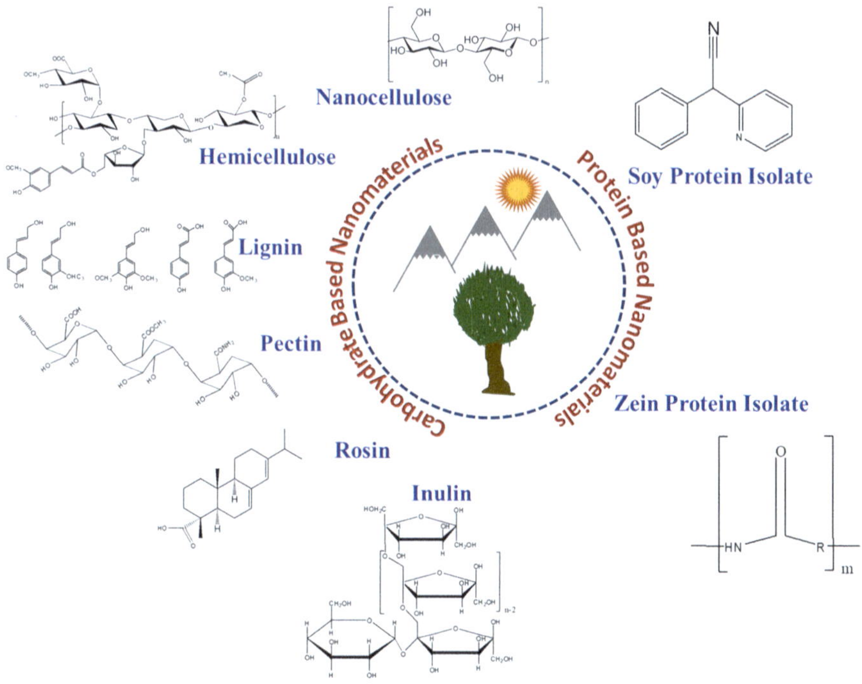

Fig. 7.1 Major structural molecular groups present in different plant-based polymeric nanomaterials

types of plant-based gum that possess charged surfaces, which can be utilized for various biomedical applications. Again, different protein-based nanomaterials, like (i) soy protein isolates is isolated from soya bean (*Glycine max L.*) and exhibits high nutritional value and desirable chemical versatilities; similarly (j) zein protein isolates is a major protein storage which constitutes (50%) protein in maize. It is a heterogeneous amphiphilic protein consisting of different peptides (α, β, γ -zein) and hydrophobic amino acids (leucine, alanine and proline). Both theses material are approved by the FDA as Generally Recognized as Safe (GRAS) for health applications. These materials have unusual properties; therefore, we need to develop these nanomaterials in the form of nanocomposite, in which complex polysaccharide structure acts as matrix, which is functionalized by either inorganic or organic nanoparticles, acting as fillers.

7.4 Different Biomedical Application of PBN

7.4.1 Controlled Drug Delivery Systems

Nanocellulose/gelatin nanocomposite in the form of cryogel with dialdehyde starch as chemical cross-linker was constructed for controlled release of 5-fluorouracil. These gels are biocompatible and have controllable porosity with reversible network structures (Li et al. 2019). Sustained co-delivery of drugs provides synergistic effects against drug resistance; thus, double layer hydrogel was constructed with cationic cellulose nanocrystals electrostatically attached with anionic alginate. This results in rapid drug release from outer layer and slow and sustained release of drug from inner layer. Thus, nanocellulose provides nonlocking and nano-obstruction effects and provides effective platform for various biomedical applications (Lin et al. 2016). Lignin is also a renewable biomacromolecule which has inherent antioxidant property and good encapsulation efficiency for poorly soluble drugs and cytotoxic agents with controlled release rates under in vitro conditions (Larraneta et al. 2018). The dentin phosphophoryn-derived peptide (DSS) lignin nanoparticles (LNPs) were developed which exhibit increased cellular uptake as compared to bare LNP in different cell lines, and this also acts as intracellular cargo delivery vehicle. These are highly biocompatible in nature, as tested in 2D and 3D spheroid tumor model (Figueiredo et al. 2019). Isolation and characterization of hemicellulose-based iron oxide functionalized nanocomposite hydrogel, in which O-acetyl-galactoglucomannan (AcGGM) was used as covalent cross linker, was found to show good potential to be utilized in different biomedical applications and regenerative medicines. It has shown sustained drug release along with controlled thermal stability, magnetization, swelling behavior and adsorption. (Zhao et al. 2015). Starch-based nanomaterials provide a new dimension to various biomedical applications. Synthesis of starch nanoparticles was carried out in ionic liquid microemulsion consisting of 1-butyl-3-methylimidazolium hexafluorophosphate ([Bmim]PF6), surfactant TX-100, 1-butanol, and water, which has shown spherical particles with slow and sustained mitoxantrone hydrochloride drug release (Zhou et al. 2014a). To increase the biological efficacy, acetylated corn starch nanoparticles were prepared by nanoprecipitation method for controlled and specific delivery of ciprofloxacin (Najafi et al. 2016). Nanomaterials based on pectin are used as encapsulating agents for various inorganic nanoparticles to reduce the toxicity concerns related to inorganic nanoparticles. Therefore, in one of the study, magnetic nanocomposites of magnetite nanoparticles, coated with 6-aminohexanoic acid and pectin, was prepared with size ~159 nm for specific delivery of hydrophobic molecular drug curcumin (Arora et al. 2017). Green chemical synthesis of nanoparticles has immense interest among researcher, as to be used in nanomedicine. Self-assembled pectin nanoparticles were prepared without using any surfactant and organic solvents, for the delivery of honokiol (hydrophobic molecule) for site-specific, slow and sustained release under in vitro conditions for the treatment of liver related diseases (Zhang et al. 2015). To reduce the side effect from chemotherapeutic drugs, pectin/chitosan nanoparticles were prepared as nanocapsules via layer-by-layer technology using silica as template. These nanocapsules exhibit high drug loading efficiency, good colloidal stability and pH-sensitive controlled doxorubicin hydrochloride release. In vitro studies suggested that these nanocapsules act as delivery vehicles which have good cytocompatibility and cellular uptake. Intravenous drug delivery with good pharmacokinetic profile is still a challenge in the field of nanomedicine. Therefore, inulin and vitamin E (INVITE) amphiphilic polymer platform was developed in the form of nanomicelles to encapsulate curcumin. It has shown improved biopharmaceutical and pharmacokinetic profiles after intravenous administration as they exhibit controlled drug release with fast cellular uptake and slow blood clearance (Tripodo et al. 2015). Another study has shown to develop an oral nanocarrier for budesonide delivery based on inulin polymer with 4-aminothiophenol (ATP) grafted onto carboxymethyl inulin with size of ~210 nm. In vitro and in vivo studies suggested its excellent potential for inflammatory bowel disease (Sun et al. 2018). Pine trees are the source of rosin and rosin-based nanoparticles have immense biomedical potential. Hydrocortisone-loaded rosin nanoparticles (167-332 nm) were prepared by dialysis method without addition of surfactant. These nanoparticles showed good cytocompatibility and pH-sensitive, slow and sustained drug release under in vitro condition (Lee et al. 2005). Soy protein-based nanoparticles provide a new insight for biomedical application. In one of the studies, nanoparticles are prepared by dispersion, desolvation, drug incorporation, cross-linking and evaporation. The nanoparticle has spherical morphology and narrow mean size distribution. Curcumin, as a model drug, was encapsulated successfully into these nanoparticles. The average size of the curcumin-loaded nanoparticles was 220–286 nm, and their zeta potential was ~ −36 mV. The highest encapsulation efficiency and loading efficiency achieved were 97.2% and 2.7%, respectively (Teng et al. 2012). To increase the targetability at the disease site, folic acid-conjugated soy protein nanoparticles were developed in aqueous solution using N-3- acrylamidophenylboronic acid monomer. These nanoparticles are ~ 200 nm in size and exhibit excellent stability under different conditions. In 2D and 3D in vitro model studies, it has shown good biological efficacy, which was also confirmed from in vivo studies (Cheng et al. 2017). Malnutrition is a major challenge in health care, and to solve this problem, nanomedicine will play a crucial role. Therefore, zein nanoparticles were developed coated with carboxymethyl chitosan for controlled release of Vitamins. These nanoparticles possess excellent physiochemical stability and improved encapsulation efficiency, which will pave a pathway to solve many nutrional deficiency diseases. Moreover, to enhance the bioavaibility of hydrophobic polyphenols, zein nanoparticles were developed to deliver resveratrol orally with enhanced bioactivity. These nanoparticles displayed a mean size diameter of 307 ± 3 nm, with a negative zeta potential (-51.1 ± 1.55 mV), and the polyphenol loading was found to be 80.2 ± 3.26 μg/mg. In vitro studies suggested that these nanoparticles have pH-independent controlled delivery of polyphenol for 48 h. The oral bioavailability of resveratrol, when administered using these nanoparticles, increased up to 50% (19.2-fold higher than the control solution of the polyphenol). Furthermore, intraperitoneal administration of the same nanoparticles, which was administered daily

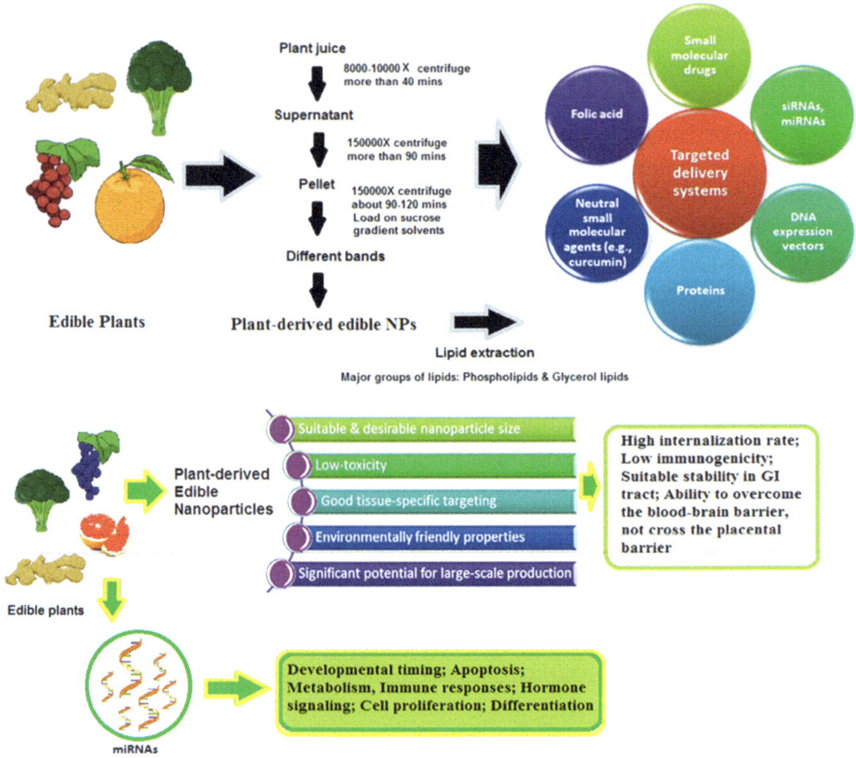

Fig. 7.2 Plant-derived edible nanoparticle for biomedical applications (Adapted from Iravani and Varma 2019, reprint with permission)

for 7 days at 15 mg/kg of body weight in mice model, were able to reduce the inflammatory response (Luo et al. 2012; Penalva et al. 2015). Gum-based nanoparticles offer a versatile role in different biomedical applications. In one of the studies, injectable acetylated gellum gum nanoparticles were developed to reduce the high antibiotic doses for skeletal-based diseases like osteomyelitis. These nanoparticles have size (258 ± 11 nm) and surface potential (−25.9 ± 0.2 mV) with site-specific sustained delivery of vancomycin. In vitro studies suggested that it has good cytocompatiblity with adjustable drug release and antimicrobial properties (Pasadowska et al. 2016). The gum-based polymeric nanoparticles were also developed to enhance bioavailability of transdermal drug diclofenac diethyl amine. These nanoparticles have shown controlled drug release evaluated by Fickian diffusion mechanism with good biocompatibility (Dias et al. 2016). In Fig. 7.2, edible plant-based nanomaterials have also been reported for targeted drug delivery (Iravani and Varma 2019). Thus, the plant-based nanomaterial provides excellent oral, transdermal and local delivery system for slow and sustained drug delivery.

7.4.2 Bioink for 3D/4D Bioprinting

3D/4D bioprinting from different renewable sources-derived nanomaterials has been used to engineer artificial tissues and organ for in vitro and in vivo biomedical studies. These nanomaterials exhibit good rheological and viscoelastic properties which protect cells from shear stress. The polysaccharide backbone also provides good biocompatibility and gelation kinetics. The 3D bioprinting utilizes inkjet and microextrusion techniques. These 3D construct biomimetic real tissues to some extent. To increase the real nature of the artificial tissue construct, 4D bioprinting has been developed, this advanced technology adds fourth dimension to the 3D printed constucts, which simulates dynamic changes of tissue in biological environment. Nanocellulose is widely explored in 3D bioink. Nanocellulose/alginate composite was prepared where the nanocellulose provides good thinning property and alginate acts as a cross-linker for bioprinting of living tissues and cells. This nanocellulose-based bioink showed non-toxic nature and cell viability of 73% and 86% after first and seventh days of 3D culture, which demonstrates the potential role of the composite in biomedical applications (Markstedt et al. 2015). Also, this bioactivated bioink hydrogel can be used for the development of drug-releasing material and wearable sensors (Leppiniemi et al. 2017). In one study, bioink composed of nanocellulose has shown increased lipid accumulation and adipogenic gene expression of adipocytes, which lead to enhanced mesenchymal stem cell maturation (Henriksson et al. 2017). Ashammakhi and co-workers developed ink comprising nanocellulose for 4D printing of tissue constructs (Ashammakhi et al. 2018). Lignin-based nanocapsules have been used for the preparation of polyvalent bioink which play a crucial role as supporting agent and activation of different polyphenol and enzymes when mixed together to impart different shades of pigmentation (Cappecchi et al. 2019). Hemicellulose, cross-linked with cellulose nanofibrils, provide a liquid bioink with excellent physiochemical properties. These bioink provide a new insight in the development of 3D bioprinting from plant-based polymeric biomaterials for regenerative medicines. In one of the studies, pectin-based nanocomposite hydrogel was used for the development of 3D biomimetic skin construct. This hydrogel utilized methacrylate pectin for the development of cell responsive bioinks with controllable biophysical and biochemical properties (Markstedt et al. 2017; Pereira et al. 2018).

7.4.3 Tissue Engineering

The basic principle of tissue engineering (TE) relies on combining living cells with a natural or synthetic support or scaffold that is biodegradable in nature to build a three-dimensional living construction which is functionally, structurally and mechanically equal or better than a tissue that is to be replaced. Nanocellulose is the most explored nanomaterial for TE because it possesses biocompatibility, water absorption, water retention, optical transparency and chemo-mechanical properties. Nanocellulose coating on 3D printed polycaprolactone (PCL) scaffold leads to

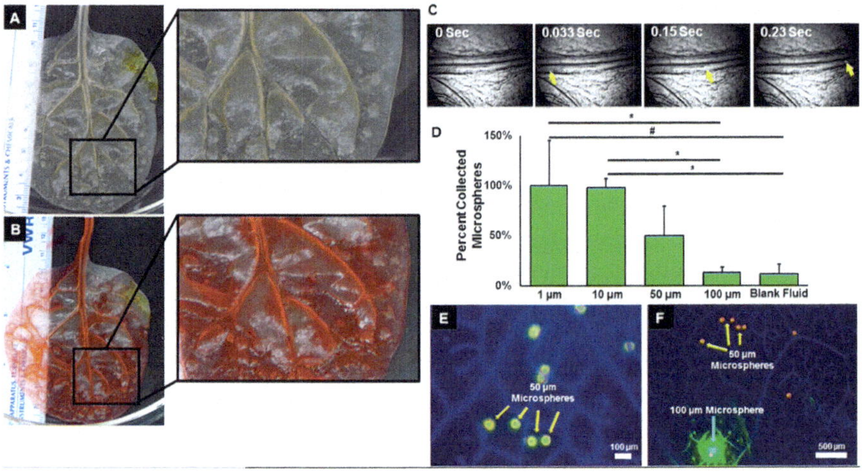

Fig. 7.3 Spinach leaf vascular scaffolds retain patency and perfusion capabilities after decellularization. Decellularized leaf **(a)** before and **(b)** after perfusion of Ponceau Red. Fluorescent microspheres of various diameters (1, 10, 50, and 100 μm) and blank fluid were perfused through a decellularized leaf and were used to determine the inner diameter flow limitations and ability of nutrient delivery. Statistical differences were found between the percentage collected between both 1 and 10 μm and the collection of 100 μm and blank fluid (∗ indicates $p < 0.05$, # indicated $p < 0.01$). **(c)** Video frames capturing microsphere location within a leaf vein over time, with the arrow tracking an individual microsphere. **(d)** Microspheres that traversed vascular walls were collected and their fluorescence quantified (n = 3). **(e, f)** Fluorescence images of leaf vasculature perfused with beads. Images show 50 and 100 μm spheres retained within the vasculature. Scale Bars: 100 μm (E), 500 μm (F). *Thus* the potency leaf vasculature helps in delivery of nutrients in complex engineered human tissues, which is still a challenge which provides excellent recellularize potential growth for human cells. (Adapted from Gershalak et al. 2017, reprint with permission)

enhanced biological efficacy for bone TE. It has shown enhanced levels of alkaline phosphatase activity, collage Type I activity and mineral formation as compared to native (PCL). It also promotes osteogenic differentiation with well-organized actin cytoskeleton and cellular proliferation (Rashad et al. 2018). Recently, immensely popular research has been conducted, which has an ability to connect animal and plant kingdoms more closely by utilizing engineering plant decellularized spinach leaves (Fig. 7.3) as a scaffold for tissue engineering (Gershlak et al. 2017). As Osteoarthritis is the major problem in which there is a high level of cartilage lesion, nanocellulose /alginate bioink was prepared printed with functional chondrocytes and human derived pluripotent stem cells which has shown a high level of hyaline-like cartilaginous tissue with elevated collagen type II (Nguyen et al. 2017). In another study, bilayer nanocomposite aerogel was developed comprising nanocellulose/polyvinyl alcohol for skin tissue engineering application (Ghafari et al. 2019). Lignin, because of its antioxidant potential, serves as favorable scaffold of neural tissue engineering (TE). The PCL/lignin nanocomposite scaffold was prepared via solvent free ring opening polymerization with excellent mechanical properties for the growth of neuron and Schwann cells (Wang et al. 2018). In another study, electrospun PCL/Lignin nanocomposite was developed as a cost-effective method for the development of biomedical scaffold. These scaffolds possess good bioactivity, morphological and mechanical property with 10% lignin content present in the scaffold. Its carbohydrate backbone mimics extracellular matrix; thus, this tissue construct will play a potential role in soft tissue engineering (Salami et al. 2017). Immunomodulatory effect was prominent for hemicellulose-based polymeric nanomaterial. Therefore, hemicellulose xylan/chitosan thermal responsive and injectable nanocomposite hydrogel was developed for bone tissue engineering. These constructs are liquid at room temperature and gel at physiological condition, and under in vivo condition, it plays a crucial role to repair large bone defects (Bush et al. 2016). The synthesis of polymeric nanomaterials from plant source leads to versatile biological properties related to tissue engineering and regenerative medicine. Therefore, pectin-based nanocomposite hydrogels were developed composed of pectin-grafted polycarpolactone and gelatin maethacryloyl along with Ca^{2+} (act as cross-linker), which provide enhanced cellular viability and mechanical property. These hydrogel-based platforms possess a high level of compressive modulus in the range of 39 kPa–5029 kPa and provide proper microenvironment and growth condition for MC3T3-E1 preosteoblasts cells (Fares et al. 2018). In another study, injectable nanocomposite hydrogel was prepared by using hyaluronic acid (HA) -adipic dihydrazide and the oligopeptide G4RGDS-grafted pectin nanofibers, in which hydrazide and aldehyde linking provide covalent interaction in this polysaccharide-based nanosystem. These cross-linkings also help to reduce toxicity and increase the mechanical properties of the hydrogel and thus can be utilized for different biomedical applications. Under in vitro condition, it was observed that oligopeptide G4RGDS concentration modulate chondrocyte bioactivity and cellular proliferation. Overall, the novel injectable multicomponent hydrogel has been proposed to be useful biomaterial scaffold for cartilage tissue regeneration (Chen et al. 2017). Similarly, pectin nanofibers were electrospun with polyhydroxybutyratuniform (PHB) via ring-opening polymerization of β-butyrolactone engineered to produce uniform nanostructure surface. This biocompatible and nontoxic scaffold showed enhanced human retinal pigmented epithelium (ARPE-19) cells proliferation, which can be utilized for retinal tissue engineering (Chan et al. 2017). Biodegradable inulin-based nanomaterials were developed to prepare amphiphilic polymer-based nanocomposites composed of Inulin and vitamin E for the delivery of hydrophobic drugs. In vivo studies on chick embryo chorioallantoic membrane showed these micelles have controlled release of drug with remarkable anti-angiogenic activity. Thus, these nanomaterials are important for biomedical applications (Mandracchia et al. 2016). Moreover, rosin nanofibers have been used for developing antimicrobial scaffold when electrospun with triethylbenzylammonium chloride (TEBAC), chitosan and silver nitrate solution. These constructs possess excellent characteristic physiochemical, mechanical strength and biological properties (Nirmala et al. 2013). In another study, gum was isolated from bael fruit for the development of nanocomposite scaffold for bone tissue engineering. This scaffold consists of bael fruit gum reinforced with nano-hydroxyapatite (n-HA) and chitosan (CS) by co-precipitation approach. This construct exhibits a wide range of antibacterial properties and biomineralization via thick apatite layer. These also possess good cell adhesion and proliferation with osteogenic differentiation, which make these biomaterials favorable for bone tissue engineering (Mirza et al. 2018). Soy protein isolate nanomaterial has also gained immense interest in tissue engineering and regenerative medicines because these are cost effective and highly abundant in nature with good biodegradability and low immunogenic response. Soy protein is incorporated with different plant-based biopolymers for enhanced biological application. In one of the applications, hydroxyethyl cellulose (HEC) and soy protein isolate (SPI) were cross-linked with epichlorohydrin to develop nanocomposite films. These films were evaluated by a series of in vitro and in vivo experiments and found to exhibit good cytocompatibility, hemocompatibility. It was found that tuning the SPI content can modulate its degeradation under in vivo conditions. Therefore, these constructs have a great potential for use in tissue engineering and regenerative medicines. Three-dimensional scaffold has been of immense interest for tissue regeneration as these provide a natural microenvironment which enhances the bioactivity and potential of cellular proliferation. Thus, protein scaffold was prepared by using soy protein as a slurry to bioprint controlled structural 3D scaffolds. The extrusion pressure and flow rate was optimized 0.0072 ± 0.0002 g/s. Addition of dithiothreitol to soy nanomaterials slurries lead to disulfide bond formation along with 1-ethyl-3-(3 dimethylaminopropyl) carbodiimide, which act as a covalent cross linker. This new nanocomposite scaffold, because of its inert and biocompatible nature, can be utilized for different tissue engineering applications (Tansaz and Boccaccini 2016; Chien et al. 2012). Another plant-based protein nanomaterial is zein. In one of the studies, nanofibrous PCL was functionalized with zein protein by electrospinning, for the fabrication of nanocomposite scaffold, which can be used for tissue engineering application. These nanomaterials can also be utilized for bone implants. In one of the studies, silica/ hydroxypropyltrimethyl ammonium chloride chitosan (HACC) /zein nanocomposite scaffolds were prepared for the cargo delivery of osteogenic factors used to repair bone defect along with this nanocomposite. It was also found that these nanocomposites possess strong antibacterial property. High mechanical strength zein-polydopamine-based nanofibrous scaffold is developed by incorporating TiO_2 metallic nanoparticles. In vitro and in vivo studies suggest that this nanomaterial has good biocompatibility and biodegradability with enhanced cellular proliferation and can be utilized for skin tissue engineering application (Jiang et al. 2007; Zhou et al. 2014; Babitha and Korrapati 2017).

7.4.4 Wound Healing

Chronic non-healing wounds represent an estimated 20 billion dollars in health care–related costs each year. Chronic wounds are characterized by delayed healing and sustained inflammation, as well as impaired extracellular matrix (ECM) function. These wounds can be caused by a number of pathologic conditions, including diabetes mellitus, vascular insufficiency, local-pressure effects, compromised nutritional and immunological states, surgeries and burns. Thus, different plant-based nanocomposite dressings can be used for the treatment of healing of these wounds. Nanocellulose can be used as wound dressing because of its inherent physiochemical and biological properties. In one of the studies, nanocellulose-reinforced polyethylene glycol (PEG) was used for wound healing as it possesses excellent mechanical property and was able to absorb exudates, which leads to healing of wounds (Sun et al. 2017). Advanced wound dressing patches of nanocellulose were developed using calcium and copper ions for tuneable, bacteriostatic and bactericidal activities. Calcium-cross-linked hydrogels inhibit growth of S. epidermidis by 266% increase in lag time, 36% increase in doubling time. Similarly, copper-cross-linked hydrogels prevented S. epidermidis growth by 49% increase in lag time and 78% increase in doubling time with bacteriostatic action towards P. aeruginosa biofilm (Basu et al. 2018a, b). Moreover, in another study, Ca^{2+} nanofibrillated hydrogel was studied for advanced wound dressing. In vitro studies suggest that this hydrogel is biocompatible in nature and reduces specific uncontrolled inflammation process during wound condition. This nanocomposite hydrogel, under in vivo condition, absorbed excessive exudates and epidermal generation (Basu et al. 2018a, b). Lignin nanofibers, because of their inbuilt structural property, can be used for wound healing applications. Gel composed of lignin nanofibers was incorporated with arginine, which act as precursor for nitric oxide release and plays an important role in wound healing. In-depth characterization showed that lignin nanofibers have 100 to 250 nm dimension and $86.28 \pm 3.50\%$ arginine was released after 24 h from the gel. Moreover, nanogel, under in vivo condition, showed increase in collagen deposition level and cell proliferation (Reesi et al. 2018). Lignin nanofiber, functionalized with chitosan nanoparticles and PVA, has shown good tensile and compressive strength with potential antibacterial property, which can be used as Band-Aid to treat chronic wounds (Zhang et al. 2019). In another study, plant-derived lignin nanoparticles, functionalized with silver nanoparticles, were utilized for the development of hydrogels. These hydrogels were based on the principle of catechol chemistry to heal chronic wounds (Gan et al. 2019). Another plant-based nanomaterial which can be used to treat wounds is hemicellulose. Nanocellulose, incorporated in hemicellulose biopolymer, result in nanocomposite hydrogels, which showed excellent physical properties. In vitro studies on fibroblast cell lines (NIH 3 T3) exhibit high levels of cell adhesion, growth and proliferation (Liu et al. 2016). Pectin is another nanomaterial which can be functionalized with chitosan nanoparticle and exhibit good physical property, viz. controllable zeta potential and stability in a wide range of pH (Birch and Schiffman 2014). In another study, nanocomposites, composed of pectin-capped silver nanoparticle of size ~8 nm, were synthesized. These nanomaterials have good potential for anti-bacterial and anti-biofilm activities, which were used for the treatment of infectious wounds (Pallavicini et al. 2017). Rosin-based nanomaterial, composed majorly of abietic acid, was used to increase MAPK kinase activity and upregulation of p38 expression responsible for wound healing and studied by western blotting. An enhanced cell migration was monitored which led to wound healing (Park et al. 2017). In another study, gum acacia biopolymer was used to formulate nanocomposite hydrogel composed of zinc oxide nanoparticles/ sodium alginate-gum acacia. This hydrogel is highly biocompatible with potential antibacterial and healing effect (Raguvaran et al. 2017). Moreover, curcumin-loaded poly(ε-caprolactone) (PCL)/gum tragacanth (GT) (PCL/GT/Cur) electrospun nanofibers construct showed wide-spectrum antibacterial activity with potential cytocompatibility, increased cells adhesion and proliferation properties. The FDA-approved soy protein nanobiomaterials also play a crucial role in wound healing. Soy proteins are engineered in the form of nanocomposite scaffold composed of cellulose acetate/ soy protein nanofibers by using rotatory jet spray technique. This construct has high water-absorption capacity which was able to clot the blood at site. When this hydrogel was treated with fibroblast cell lines, it has shown increased proliferation and β1 integrin expression. Under in vivo condition, these nano-scaffolds were used to re-epithelize the damaged tissue (Ahn et al. 2018). In another study, polyurethane and soy protein nanofiber were electrospun with zinc oxide nanoparticles to produce fibrous nanomats. In-depth physical characterization results ~300–350 nm size fibers with smooth nanotopography; with increased biological efficacy for wound healing (Kumar et al. 2017). In addition, in vivo studies of electrospun soy protein nanofibers were carried out on porcine model of wound healing, which results in similar biological response of this construct compared to market Tegaderm® mats (Gerstenhaber et al. 2014). Zein-based nanomaterial can also be used for wound healing because of their biocompatible nature. In one of the studies, zein/collagen protein was co-electrospun with acetic acid solutions to develop fibrous nanomats. These nanomats also showed controlled drug release. In vitro studies showed increased cell viability and proliferation and in vivo results exhibit enhanced wound healing capacity (Lin et al. 2012). In an another study, electrospinning was utilized for poly(l-lactide)(PLLA)/zein nanofiber mats loaded with Rana chensinensis skin peptides (RCSPs) for wound healing application (Zhang et al. 2016a, b). Moreover, zein nanomats of size ~350–500 nm, functionalized with silver nanoparticles (~20 nm), has resulted in good cytocompatibility and antibacterial activity. Moreover, these nanomats have shown a high level of wound-healing activity (Dashdorj et al. 2015).

7.4.5 Antimicrobial Application

The over-prescription, abuse and misapplication of antibiotics have led to an escalating drug resistance in pathogenic microorganisms, which is associated with increased morbidity and mortality. Exacerbating antibiotic resistance problem relies on the fact that only a few new families of antibiotics have been introduced in clinics since 1960. Thus, antibiotic resistance was estimated to kill 700,000 people per year worldwide, and this number was predicted to increase to 10 million by 2050. Misuse of antibiotics over time has led to the development of antimicrobial resistance, which is now a global concern. The surface functionalization of nanocellulose to combat microbial resistance led to a new avenue of antimicrobial material. In one of the studies, nanocellulose, functionalized with lysozyme and nisin, was synthesized and characterized using FT-IR and XPS, which has shown the proper cross-linking of niacin and lysozyme on nanocellulose surface. The efficacy of the developed nanocomposite against gram positive and gram-negative bacteria showed increased killing of bacteria and inhibition of biofilm (Tavakolian et al. 2018). In another study, antimicrobial surface was developed using nanocellulose. For these, nanocellulose, functionalized with porphyrin, was utilized as a cross-linker. This process was catalyzed by Cu(I)-catalyzed azide-alkaline Huisgen 1,3-dipolar cyclo-addition, leading to covalent bond formation between porphyrin alkynes and azide group on cellulose surface. These nanocomposites have good dispersibility in the aqueous solution, and also showed good photodynamic inactivation against a wide spectrum of bacterial growth (Feese et al. 2011) Also, cellulose nanocrystal, functionalized with polyrhodanine using ferric chloride as initiator and oxidant, was reported to possess high antibacterial potential (Tang et al. 2015). Lignin-based nanomaterials were also explored for their antibacterial property because of their inherent polyphenolic structure and antioxidant nature. Silver ions-infused lignin nanoparticles were developed, which were coated with cationic polyelectrolyte. These showed a wide spectrum killing of antimicrobial resistance bacterial strains under in vitro and in vivo conditions (Richter et al. 2015). In one of the studies, conductive lignin nanocomposite films were developed by surface functionalization with gentamicin and iron oxide nanoparticles via matrix-assisted pulsed laser evaporation technique. These nanofilms have shown a high level of biocompatibility and also reduced and inhibited the growth of anti-microbial strains (Visan et al. 2019). Pectin-based plant nanomaterial is also utilized to tackle antimicrobial resistance using pectin/zirconium(IV) selenotungstophosphate. This material possesses excellent mechanical and thermal stability and also retarded the growth of *Staphylococcus aureus* (Sharma et al. 2014). In another study, pectin biopolymer, incorporated with germanium dioxide nanocomposite, was prepared by hydrothermal method. The result showed that ~ 70 nm germanium dioxide nanoparticles were functionalized on pectin nanofibers, which were confirmed by FT-IR studies, and these nanocomposites possess good thermal stability and tensile strength. Also, this nanomaterial exhibits good antimicrobial property and can be utilized for biomedical application (Chauhan et al. 2017). Inulin-based nanocomplex, composed of gold glyconanoparticles and Listeria monocytogenes Listeriolysin O (LLO91–99) peptide, induces a high level of immune response against listeria infection by enhancing T cell proliferation. This nanocomposite is biocompatible and can be used for the development of vaccine against pathogens (Rodriguez et al. 2015). In another study, plant-based zein polymer is used as capping agent for magnesium oxide nanoparticles synthesis, which has shown increased bioavailability and bioactivity by preventing nanoparticle aggregation. This study has shown antibacterial and antifungal activities validated by the Kirby-Bauer test, and a modified direct contact test. The results lead to improvement of dental material and can improve oral health (Naguib et al. 2018). In another study, biodegradable zein/chitosan/cinnamon essential oil nanocomposite films were synthesized, which resulted in good physical and mechanical properties with enhanced antibacterial action (Vahedikia et al. 2019).

7.4.6 Nanomaterials as Anti-proliferative Agent Against Cancer Cells

Drug-resistant cancerous disease has emerged as a new threat, where chemotherapeutic resistance has failed to treat cancer. In this respect, new plant-based nanomaterial can provide new opportunities for the treatment of these diseases because of their enhanced bioactivity. In one of the studies, cellulose nanocrystals, functionalized with curcumin and cyclodextrin by ionic interaction, led to controlled delivery of curcumin under in vitro prostate and colorectal cancer cell treatment. These nanocomplexes have resulted in new anti-cancer drug delivery systems (Ntoutoume et al. 2016). Usual failure of nanomaterial drug at phase II clinical trial resulted due to the inability of the nanomaterials to bypass P-glycoprotein efflux pump. Thus, magnetic nanocapsule-reinforced protamine carboxymethyl nanocellulose delivery nanomaterial was developed, which can bypass all the cellular barriers and site-specific drug delivery can be achieved (Elumalai et al. 2015). In another study, nanocellulose attached with Fe_3O_4-Ag_2O quantum dots conjugated with Etoposide and Methotrexate molecular drugs has shown good cytotoxic potential and can be utilized for skin cancer treatment (Fakhri et al. 2017). Also, graphene quantum dot functionalized carboxymethyl-nanocellulose loaded with doxorubicin has shown pH-dependent controlled drug release with wide-spectrum anti-cancerous activity (Javanbakht and Namazi 2018). Lignin-based biopolymer has also been explored for anti-cancerous studies and it showed narrow size distribution and good stability in physiological media. These nanoparticles can be used to encapsulate water insoluble drugs with pH-dependent, slow and sustained characteristic release. It also exhibits good cellular viability to normal cell lines and cytotoxic activity against cancer cell lines (Figueiredo et al. 2017). Moreover, multifunctional lignin nanoparticle was developed for the treatment of drug resistant colon cancer. These nanoparticles were encapsulated with anti-cancerous drugs, functionalized with CD44 receptors, which lead to the site-specific drug targeting towards the cancerous tissues (Siddiqui et al. 2018). Pectin-based nanoparticles were also evaluated for anti-cancerous potential. Methotrexate carbodiimide-conjugated pectin nanoparticles were prepared by using an ionotropic gelation having uniform size. These nanoparticles have showed sustained drug delivery at pH 7.4 towards HepG2 cancerous cell lines and thus can be used as nanodrug (Chittasupho et al. 2013). Inulin-based nanomaterial is explored for potential anti-cancerous activity. Thus, gold nanosphere, capped with thiolated polyethyleneglycol and inulin-amino derivative, was synthesized having good stealth and physical properties. In vitro studies on human cervical (Hela) and lung (A549) cancer cell lines, along with the coculture (cancer/noncancer) model, suggest that these nanoparticles were internalized and resulted in controlled delivery of drug inside the tumor, which was confirmed with fluorescence study (Volsi et al. 2016). Also, in another study, gold nanoparticle of size 40 nm was capped with inulin-folate receptor and showed enhanced uptake in breast cancer cell lines as compared to normal epithelial cell lines thus confirmed the specific anti-cancerous potential (Licciardi et al. 2016). Soy-based protein nanomaterial, attached with folic acid cargo vehicle, was developed for site-specific delivery and to enhance intake capability of drugs against cancer tissue. This nanoparticle has shown good biological and physical properties along with better stability of nanoparticle with no nanoparticle aggregation. It possesses 96.7% encapsulation and 23% loading efficiencies, respectively. In vitro spheroid tumor model also suggested that this nanoparticle documented high level of killing against tumor growth (Yao et al. 2017). Zein-based nanomaterial, which was synthesized by various top down approaches, is utilized for anti-cancerous studies. Paclitaxel-encapsulated zein nanoparticles were prepared and exhibited enhanced bioavailability and stability properties. However, under in vitro conditions, these nanomaterials showed cytotoxic nature. Gastric cancer has emerged as a major threat to human population. This led to the development of core shell zein nanocomposite, which was composed of sorafenib-loaded zein/chondroitin sulphate nanoparticles. This nanoparticle possesses spherical morphology and greater than 90% encapsulation efficiency with sustained release for 120 h. These nanoparticles were easily taken by gastric cancer cell lines (SGC7901) and showed enhanced apoptosis by activation of caspases 3/7 (Elzoghby et al. 2017; Gagliardi et al. 2019; Cheng et al. 2016).

7.4.7 Molecular Bioimaging

Plant-based nanomaterials can be used for next generation biomedical imaging technology. Nanocellulose, functionalized with graphene quantum dot, was used as injectable fluorescent hydrogel (Khabibullin et al. 2017). In another study, carbon quantum dots (prepared by microwave synthesis method) functionalized on 2,2,6,6-tetramethylpiperidine-1-oxyl radical (TEMPO) oxidized nanocellulose (prepared by hydrochloric acid hydrolysis) were synthesized and characterized by FT-IR and XPS studies. In vitro studies suggested that the nanocomposite can easily be internalized by macrophages and cancer cell lines and showed good fluorescence response (Guo et al. 2017). Also, carboxymethylcellulose (CMC)/ZnCdS quantum dot was reported for molecular bioimaging of cancer cell (Mansur et al. 2017). In another study, wood lignocellulosic biomass was functionalized with rhodamine B as a pH-sensing probe for cancer cell bioimaging. These nanocomplexes were also specifically cytotoxic to cancer cell lines only (Xue et al. 2016). Pectin-based natural biopolymer-derived nanomaterials are also used to develop carbon dots. Citrus pectin-based hydrogel, when heated with NaOH at 180 °C, has shown small sized (~2.5 nm) carbon dots with good photostability and cyan fluorescence. Therefore, it can be utilized for fluorescent bioimaging (Zhao et al. 2014). Also, soy protein-encapsulated 2D MoS_2 (SP-MoS_2) nanosheets were developed which has shown excellent bioavailability and photothermal effect on breast tumor cell lines with bioimaging ability and was able to inhibit the tumor growth (Li et al. 2016). Zein-based nanoparticles, encapsulated with 5-fluorouracil, were conjugated with quantum dots and have shown controlled drug delivery and cytotoxic nature against L929 and MCF-7 cell lines with proper cellular bioimaging (Aswathy et al. 2012).

7.4.8 Protein Aggregation-Related Diseases

There are various physiological and neurodegenerative diseases which can be caused by protein aggregation. Different nanomaterials and small molecules have been reported for protein inhibition, but plant-based polymeric nanomaterial is the least explored. In one of the studies, nanocellulose functionalized with iron oxide nanoparticles led to organic and inorganic type of hybrid nanomaterials, which can provide simultaneous hydrophobic and hydrophilic interaction to inhibit glycation-induced protein aggregation (Singla et al. 2019a, b).

7.4.9 Other Biomedical Application

Nanocellulose has been utilized for the delivery of human adipose tissue mesenchymal stem cells (ATMS) for the reduction of inflammation and surgical wounds. Nanocellulose functionalized with glutaraldehyde has shown high mechanical and biological properties of adhesion, cell migration and proliferation, and under ex vivo condition, it has shown good cell viability of ATMS (Mertaniemi et al. 2016). Therefore, this type of nano-biomaterial can be utilized as post-operative surgical material. Nanocellulose can be engineered in different ways for various biomedical applications because of their inert physiochemical properties (Curvello et al. 2019). Lignin has been used in the form of nanocontainer via converting it into phenoxy ether for the encapsulation of dye. These have been utilized as an image contrasting agent with enhanced bioimaging application (Cauley and Wilson 2017). Also, nanocomposite of quaternized hemicellulose films have been reported, which was functionalized with montmorillonite onto which polyvinyl alcohol (PVA) and chitin nanowhiskers were used as nanofillers. Proper characterization of this type of film suggested that PVA acts as better filler and imparts excellent mechanical properties as compared to chitosan nanowhisker-based films. It was proposed that these nanocomposite-based films can be used as coating and packing material for various biomedical systems (Chen et al. 2015). Pectin-based biomatrix, cross-linked with cerium oxide nanoparticles, has also been used as a good biomedical material which was utilized for different diseases because of their wide photoluminescence emission spectra, colloidal stability, antimicrobial efficacy and lower toxicity (Patil et al. 2016). Pectin-based nanofibers are also used for encapsulation of probiotic bacteria, which are resistant to gastrointestinal degradation and lead to better biological activity by overcoming lactose intolerance (Khorasani and Shojaosadati 2017). Ethylenediamine cross-linked inulin-based polymeric nanomaterials are also used as the coating material for superparamagnetic iron oxide nanoparticles (SPIONS), which has reduced toxicity and aggregation of nanoparticles. Also, these nanoparticles were complexed with siRNA, which were able to reduce tumor cellular viability (Licciardi et al. 2015). Rosin-based polymeric nanoparticle is also used, which has shown fast precipitation in acidic medium, but no aggregation was reported for alkali medium. These nanocomposites can be used for versatile biomedical application (Kozlova et al. 2017). Soy-based nanoparticles when incorporated with rice bran oil lead to nano-emulsion formation and lead to good antioxidant, antimicrobial and proteolytic activity, which can be utilized as functional non-dairy yogurt. Moreover, soy protein isolate, attached with bovine serum albumin, can be used for inulinase enzyme immobilization (Sengupta et al. 2019; Mikani et al. 2018). Zein protein nanoparticle has been utilized and evaluated for oral drug delivery under in vivo *C. elegans* model (Lucio et al. 2017).

7.5 Requirement of Plant-Based Nanomaterial from a Clinical Perspective

Many unanswered questions about the fate of the material are still to be addressed to use plant-based nanomaterial for biomedical applications. These are as follows:

(a) Nanomaterial physiochemical and biological properties under complex environment of culture condition still need to be evaluated.
(b) Proper animal model to study the effect of plant-based nanomaterial under different conditions and disease states need to be developed.
(c) Easy and specific route of nanomaterial delivery should be identified.
(d) Appropriate degradation profile of the nanomaterial should be known and its accumulation should not affect the metabolism of body.
(e) Material should be non-immunogenic and non-toxic in nature.

Animal model studies of nanocellulose have been widely explored for different biomedical application because of its excellent physiochemical and biological properties. Thus, proper animal studies have to be carried out for its evaluation of toxicity. Yanamala et al. studied two different forms of nanocellulose, that is, cellulose nanocrystals and cellulose nano-powder, on C57BL/6 mice and the results suggested that size of the nanomaterial is the major criterion for the upregulation of cytokines response and inflammation (Yanamala et al. 2014). In another study, nanocellulose has shown non-hazardous nature when taken as 1%w/w form of dosage for in vivo study in rat model and showed absence of immune-histological toxic markers in any of the organs. Also, lignin nanoparticles, encapsulated with resveratrol, have shown specific delivery and anti -tumor effects with low toxicity and immunogenic response. The nanomaterials obtained from plants are highly biocompatible in most of the reports discussed in previous sections, but the mechanistic insight at the disease site and its exact clearance has not been studied in detail. Therefore, functionalized plant-based nanomaterial should be designed in such a way for clinical use that its physiochemical and mechanical properties of the nanomaterials should be retained with proper biodegradation profiles. In Table 7.1,

Table 7.1 Some plant-based materials used as clinical trial for different biomedical applications

Plant-based composite	Clinical application	Conclusion	References
Nanocellulose	Wound dressing with potential to be used during burns having skin graft treatment	a. Nanofibrillated cellulose (NFC) of different thickness have been used as wound dressing for burns with skin donor site treatment. b. Clinical trial was conducted on nine burn patients and nanocellulose exhibits better result than commercial dressing Suprathel® (PMIPolymedics, Germany). c. the developed dressing dehydrated and attached itself to donor site during the first days. After 11–21 days, when healthy epithelization of skin has developed, the dressing has detached from the burn site. Suprathel®, as compared to NFC dressing, takes 16–28 days for detachment from the wound bed. d. NFC dressings are highly biocompatible, no allergic and inflammatory response, easily attaches to the wound until the skin graft donor burn site is renewed. It also detaches from the reephithilized skin by itself.	Hakkarainen et al. (2016)
Lignin loaded with bioactive ascorbic acid	Treatment for herpes Siplex virus	a. Lignin loaded with ascorbic acid as a highly antioxidant nanocomposite, which is clinically evaluated for the treatment of herpes simplex virus type (1). b. Clinically, pilot study was carried out for both the genders, between 4 and 61 year old, in which odromic (16 patients), erythema (11 patients), papule edema (1 patient), vesicle/ pustule (13 patients) and ulcer stages (7 patients), respectively. c. the patients who began the lignin-ascorbic acid treatment within the first 48 h of symptom onset did not develop HSV-1, as compared with previous episodes.	Lopez et al. (2009)

(continued)

Table 7.1 (continued)

Plant-based composite	Clinical application	Conclusion	References
Hemicelluose / vitamins conjugates	Enhanced immune system function	a. Hemicellulose-based Natramune (PDS-2865) nanobiotechnological product stimulates phagocytosis (65%), nitric oxide production against (517%) by macrophages cells against pathogens and prome cell viability (155%) under inflammatory condition. b. Natramune (PDS-2865) enhanced human circulating total lymphocyte levels (18%) during bacterial infection.	Weeks and Perez. (2009)
O acetylated pectin	Vaccine for typhoid	a. Clinical phase I trial, based on pectin conjugated vaccines, has been developed for increased efficacious delivery of vaccines against typhoid. b. Pectin conjugated with a multivalent recombinant exoprotein A from Pseudomonas aeruginosa formed the composite (OAcPec-rEPA). c. A nanoscience-based plant typhoid vaccine, is safe and immunogenic in adult volunteers.	Szu et al. (2014)
Inulin	Treatment of type 2 diabetes	a. Clinical trial was based on the study of inulin-based nanocomposite and its effect on type 2 diabetes patients. b. Controlled feeding efficacy and release was monitored on 48 adults (40–75 yrs) with prediabetes or a score $\geq$ 5 on the American Diabetes Association (ADA) risk screener to 6 weeks. c. Composite as prebiotic supplementation will increase cardiometabolic health.	Mitchell et al. (2015)
Zein encapsulated vardenafil	Cardiovascular related disorder	a. Zein/alpha liponic acid nanosphere encapsulated with vardenafil for improved drug delivery and bioavailability. b. this material exhibits good physiochemical and biological properties under in vitro and	Ahmed (2018)

(continued)

Table 7.1 (continued)

Plant-based composite	Clinical application	Conclusion	References
		in vivo conditions with excellent clinical pharmacokinetic profiles. c. Showed better efficacy of vardenafil, with 2.5 times higher bioavailability in healthy volunteer.	
Hydrolysed guar gum loaded Rifaximin	Treatment of small intestinal bacterial overgrowth (SIBO)	a. this nanocomposite is used to increase intestinal motility, which decreases SIBO. b. Eradication rate of SIBO was 62.1% in the rifaximin group (both on per-protocol and intention-to-treat analyses), and 87.1% (per-protocol, $P = 0.017$) and 85.0% (intention-to-treat, $P = 0.036$) in the rifaximin/hydrolysed guar gum group. c. Clinical improvement was observed in 86.9% and 91.1% of eradicated cases in rifaximin and rifaximin/guar gum nanocomposite, respectively ($P = 0.677$).	Furnari et al. (2010)

some plant-based materials have been described which are in clinical trial application for different biomedical applications, showing positive effect of research and development on these biomaterials.

Clinical trial design of plant-based nanomaterials has been immensely studied for different biomedical applications because of their advanced mechanical, chemical and biological properties. But there are only few marketed and clinically used materials available. Most of these nanomaterials, because of their unexplored clearance mechanism and their unevaluated effect inside the body for prolonged duration, may lead to undetermined biological responses which will result in reduced clinical outcome. Hakkarainen et al. reported that patients treated with nano-fibrillated cellulose, as skin graft donor nanomaterial, have shown greater epithelialization, less pain and irritation as compared to commercial Suprathel® (Hakkarainen et al. 2016). Thus, plant-based nanomaterial has shown advanced biological efficacy, though their long-term clinical trial need to be thoroughly understood. Proper understanding of the disease has to be determined with in-depth mechanistic insight of these tuned nanomaterials for different biomedical application. Different techniques have to be developed to access the biomechanical, physical and biological property of engineered nanomaterials. Moreover, long-term follow-up is required to validate plant-based nanomaterial as a first line of treatment for different biological diseases.

7.6 Conclusion and Future Perspective

Plants-based nanomaterials need to be used in biomedical application, but requires to include several factors, such as cell response towards nanocomposites and their clearance mechanism from the biological systems. Also, isolation and characterization of these nanomaterials are also important aspects which have to be covered in detail to utilize these nanomaterials for biological application. These biomaterials, in the form of nanocomposites, possess good biocompatibility, biodegradability, porosity and mechanical stability with enhanced efficacy towards disease sites. PBNs impart the versatile platform, as these act as a suitable nanomaterial to be used as good cross-linking agent due to the presence of poly chemical functionalities and can be converted to hydrogel, aerogel and injectable nanocomposite materials. Incorporation of the different bioactive molecules and inorganic nanomaterial has also imparted good biological properties for these nanosystems, which also helps to resolve the solubility issues related to these nanomaterials. Since most of the activities have been short term and conducted on animal model, therefore, potential of these different types of nanomaterial should be reported to clinical trials. The main focus of the present chapter is to evaluate biological and mechanical properties of these developed nanomaterials under biological microenvironment. The ultimate success of these nanomaterials is to engineer these in accordance to the biomedical application with proper long term in vivo studies.

Acknowledgements The authors would like to thank Director, CSIR- IHBT for his constant support and encouragement. AA acknowledges the financial support from CSIR (MLP-201) and DST (GAP-0214; EMR/2016/003027). SMSA and AID thank DBT and DST, respectively, for providing SRF. Academy of Scientific and Innovative Research, New Delhi is duly acknowledged. The CSIR-IHBT communication number of this manuscript is 4499.

References

Ahmed OA (2018) Development and single dose clinical pharmacokinetics investigation of novel zein assisted-alpha lipoic acid nanoencapsulation of vardenafil. Sci Rep 8(1):15802

Ahn S, Chantre CO, Gannon AR, Lind JU, Campbell PH, Grevesse T, O'Connor BB, Parker KK (2018) Soy protein/cellulose nanofiber scaffolds mimicking skin extracellular matrix for enhanced wound healing. Adv Healthc Mater 7(9):1701175–1701188

Arora V, Sood A, Shah J, Kotnala RK, Jain TK (2017) Synthesis and characterization of pectin-6-aminohexanoic acid-magnetite nanoparticles for drug delivery. Mater Sci Eng C 80:243–251

Ashammakhi N, Ahadian S, Zengjie F, Suthiwanich K, Lorestani F, Orive G, Ostrovidov S, Khademhosseini A (2018) Advances and future perspectives in 4D bioprinting. Biotechnol J 13(12):1800148–1800160

Aswathy RG, Sivakumar B, Brahatheeswaran D, Fukuda T, Yoshida Y, Maekawa T, Kumar DS (2012) Biocompatible fluorescent zein nanoparticles for simultaneous bioimaging and drug delivery application. ANSN 3(2):025006–025013

Babitha S, Korrapati PS (2017) Biodegradable zein–polydopamine polymeric scaffold impregnated with TiO2 nanoparticles for skin tissue engineering. Biomed Mater 12(5):055008–550020

Basu A, Heitz K, Stromme M, Welch K, Ferraz N (2018a) Ion-crosslinked wood-derived nanocellulose hydrogels with tunable antibacterial properties: Candidate materials for advanced wound care applications. Carbohydr Polym 181:345–350

Basu A, Celma G, Stromme M, Ferraz N (2018b) In vitro and in vivo evaluation of the wound healing properties of nanofibrillated cellulose hydrogels. ACS Appl Bio Mater 1(6):1853–1863

Birch NP, Schiffman JD (2014) Characterization of self-assembled polyelectrolyte complex nanoparticles formed from chitosan and pectin. Langmuir 30(12):3441–3447

Bush JR, Liang H, Dickinson M, Botchwey EA (2016) Xylan hemicellulose improves chitosan hydrogel for bone tissue regeneration. Polym Adv Technol 27(8):1050–1055

Capecchi E, Piccinino D, Bizzarri BM, Avitabile D, Pelosi C, Colantonio C, Calabro G, Saladino R (2019) Enzyme-lignin Nanocapsules are sustainable catalysts and vehicles for the preparation of unique polyvalent bioinks. Biomacromolecules 20(5):1975–1988

Cauley AN, Wilson JN (2017) Functionalized lignin biomaterials for enhancing optical properties and cellular interactions of dyes. Biomater Sci 5(10):2114–2121

Chan SY, Chan BQY, Liu Z, Parikh BH, Zhang K, Lin Q, Su X, Kai D, Choo WS, Young DJ, Loh XJ (2017) Electrospun pectin-polyhydroxybutyrate nanofibers for retinal tissue engineering. ACS Omega 2(12):8959–8968

Chauhan NPS, Gholipourmalekabadi M, Mozafari M (2017) Fabrication of newly developed pectin–GeO_2 nanocomposite using extreme biomimetics route and its antibacterial activities. J Macromol Sci A 54(10):655–661

Chen GG, Qi XM, Li MP, Guan Y, Bian J, Peng F, Yao CL, Sun RC (2015) Hemicelluloses/montmorillonite hybrid films with improved mechanical and barrier properties. Sci Rep 5:16405

Chen F, Ni Y, Liu B, Zhou T, Yu C, Su Y, Zhu X, Yu X, Zhou Y (2017) Self-crosslinking and injectable hyaluronic acid/RGD-functionalized pectin hydrogel for cartilage tissue engineering. Carbohydr Polym 166:31–44

Cheng HT, Hsieh SY, Chen TH, Hung PF, Pan SH (2016) Sorafenib-fortified zein–chondroitin sulphate biopolymer nanoparticles as a novel therapeutic system in gastric cancer treatment. RSC Adv 6(62):57266–57274

Cheng X, Wang X, Cao Z, Yao W, Wang J, Tang R (2017) Folic acid-modified soy protein nanoparticles for enhanced targeting and inhibitory. Mater Sci Eng C 71:298–307

Chien KB, Makridakis E, Shah RN (2012) Three-dimensional printing of soy protein scaffolds for tissue regeneration. Tissue Eng Part C Methods 19(6):417–426

Chittasupho C, Jaturanpinyo M, Mangmool S (2013) Pectin nanoparticle enhances cytotoxicity of methotrexate against hepG2 cells. Drug Deliv 20(1):1–9

Curvello R, Raghuwanshi VS, Garnier G (2019) Engineering nanocellulose hydrogels for biomedical applications. Adv Colloid Interface Sci 267:47–61

Dashdorj U, Reyes MK, Unnithan AR, Tiwari AP, Tumurbaatar B, Park CH, Kim CS (2015) Fabrication and characterization of electrospun zein/Ag nanocomposite mats for wound dressing applications. Int J Biol Macromol 80:1–7

Dias SFL, Nogueira SS, De França DF, Guimaraes MA, De Oliveira Pitombeira NA, Gobbo GG, Primo FL, De Paula RCM, Feitosa JPA, Tedesco AC, LCC N (2016) Acetylated cashew gum-based nanoparticles for transdermal delivery of diclofenac diethyl amine. Carbohydr Polym 143:254–261

Elumalai R, Patil S, Maliyakkal N, Rangarajan A, Kondaiah P, Raichur AM (2015) Protamine-carboxymethyl cellulose magnetic nanocapsules for enhanced delivery of anticancer drugs against drug resistant cancers. Nanomedicine: NBM 11(4):969–981

Elzoghby A, Freag M, Mamdouh H, Elkhodairy K (2017) Zein-based nanocarriers as potential natural alternatives for drug and gene delivery: focus on cancer therapy. Curr Pharm Des 23 (35):5261–5271

Fakhri A, Tahami S, Nejad PA (2017) Preparation and characterization of Fe_3O_4-Ag_2O quantum dots decorated cellulose nanofibers as a carrier of anticancer drugs for skin cancer. J Photoch Photobio B 175:83–88

Fares MM, Sani ES, Lara RP, Oliveira RB, Khademhosseini A, Annabi N (2018) Interpenetrating network gelatin methacryloyl (GelMA) and pectin-g-PCL hydrogels with tunable properties for tissue engineering. Biomater Sci 6(11):2938–2950

Feese E, Sadeghifar H, Gracz HS, Argyropoulos DS, Ghiladi RA (2011) Photobactericidal porphyrin-cellulose nanocrystals: synthesis, characterization, and antimicrobial properties. Biomacromolecules 12(10):3528–3539

Figueiredo P, Ferro C, Kemell M, Liu Z, Kiriazis A, Lintinen K, Florindo HF, Yli-Kauhaluoma J, Hirvonen J, Kostiainen MA, Santos HA (2017) Functionalization of carboxylated lignin nanoparticles for targeted and pH-responsive delivery of anticancer drugs. Nanomedicine 12 (21):2581–2596

Figueiredo P, Sipponen MH, Lintinen K, Correia A, Kiriazis A, Yli-Kauhaluoma J, Osterberg M, George A, Hirvonen J, Kostiainen MA, Santos HA (2019) Preparation and characterization of dentin Phosphophoryn-derived peptide-functionalized lignin nanoparticles for enhanced cellular uptake. Small 15(24):1901427

Furnari M, Parodi A, Gemignani L, Giannini EG, Marenco S, Savarino E, Assandri L, Fazio V, Bonfanti D, Inferrera S, Savarino V (2010) Clinical trial: the combination of rifaximin with partially hydrolysed guar gum is more effective than rifaximin alone in eradicating small intestinal bacterial overgrowth. Aliment Pharmacol Ther 32(8):1000–1006

Gagliardi A, Bonacci S, Paolino D, Celia C, Procopio A, Fresta M, Cosco D (2019) Paclitaxel-loaded sodium deoxycholate-stabilized zein nanoparticles: characterization and in vitro cytotoxicity. Heliyon 5(9):e02422–e02432

Gan D, Xing W, Jiang L, Fang J, Zhao C, Ren F, Fang L, Wang K, Lu X (2019) Plant-inspired adhesive and tough hydrogel based on Ag-lignin nanoparticles-triggered dynamic redox catechol chemistry. Nat Commun 10(1):1487–1497

Gershlak JR, Hernandez S, Fontana G, Perreault LR, Hansen KJ, Larson SA, Binder BY, Dolivo DM, Yang T, Dominko T, Rolle MW (2017) Crossing kingdoms: using decellularized plants as perfusable tissue engineering scaffolds. Biomaterials 125:13–22

Gerstenhaber JA, Brodsky R, Huneke RB, Lelkes PI (2014) Electrospun soy protein scaffolds as wound dressings: Enhanced re-epithelialization in a porcine model of wound healing. Wound Medicine 5:9–15

Ghafari R, Jonoobi M, Amirabad LM, Oksman K, Taheri AR (2019) Fabrication and characterization of novel bilayer scaffold from nanocellulose based aerogel for skin tissue engineering applications. Int J Biol Macromol 136:796–803

Guo J, Liu D, Filpponen I, Johansson LS, Malho JM, Quraishi S, Liebner F, Santos HA, Rojas OJ (2017) Photoluminescent hybrids of cellulose nanocrystals and carbon quantum dots as cytocompatible probes for in vitro bioimaging. Biomacromolecules 18(7):2045–2055

Hakkarainen T, Koivuniemi R, Kosonen M, Escobedo-Lucea C, Sanz-Garcia A, Vuola J, Valtonen J, Tammela P, Makitie A, Luukko K, Yliperttula M (2016) Nanofibrillar cellulose wound dressing in skin graft donor site treatment. JCR 244:292–301

Henriksson I, Gatenholm P, Hagg DA (2017) Increased lipid accumulation and adipogenic gene expression of adipocytes in 3D bioprinted nanocellulose scaffolds. Biofabrication 9 (1):015022–015032

Iravani S, Varma RS (2019) Plant-derived edible nanoparticles and miRNAs: emerging frontier for therapeutics and targeted drug-delivery. ACS Sustain Chem Eng 7(9):8055–8069

Javanbakht S, Namazi H (2018) Doxorubicin loaded carboxymethyl cellulose/graphene quantum dot nanocomposite hydrogel films as a potential anticancer drug delivery system. Mater Sci Eng C 87:50–59

Jiang H, Zhao P, Zhu K (2007) Fabrication and characterization of zein-based nanofibrous scaffolds by an electrospinning method. Macromol Biosci 7(4):517–525

Khabibullin A, Alizadehgiashi M, Khuu N, Prince E, Tebbe M, Kumacheva E (2017) Injectable shear-thinning fluorescent hydrogel formed by cellulose nanocrystals and graphene quantum dots. Langmuir 33(43):12344–12350

Khorasani AC, Shojaosadati SA (2017) Pectin-non-starch nanofibers biocomposites as novel gastrointestinal-resistant prebiotics. Int J Biol Macromol 94(Part A):131–144

Kozlova EA, Atkin VS, Zakharevich AM, Markin AV, Skaptsov AA (2017) Synthesis and investigation of rosin nanoparticles. In Saratov Fall Meeting 2016: optical technologies in biophysics and medicine XVIII (vol 10336). International Society for Optics and Photonics, p 1033615

Kumar NS, Santhosh C, Sudakaran SV, Deb A, Raghavan V, Venugopal V, Bhatnagar A, Bhat S, Andrews NG (2017) Electrospun polyurethane and soy protein nanofibres for wound dressing applications. IET Nanobiotechnol 12(2):94–98

Larranneta E, Imizcoz M, Toh JX, Irwin NJ, Ripolin A, Perminova A, Dominguez-Robles J, Rodríguez A, Donnelly RF (2018) Synthesis and characterization of lignin hydrogels for potential applications as drug eluting antimicrobial coatings for medical materials. ACS Sustain Chem Eng 6(7):9037–9046

Lee CM, Lim S, Kim GY, Kim DW, Rhee JH, Lee KY (2005) Rosin nanoparticles as a drug delivery carrier for the controlled release of hydrocortisone. Biotechnol Lett 27(19):1487–1490

Leppiniemi J, Lahtinen P, Paajanen A, Mahlberg R, Metsaa Kortelainen S, Pinomaa T, Pajari H, Vikholm Lundin I, Pursula P, Hytonen VP (2017) 3D-printable bioactivated nanocellulose–alginate hydrogels. ACS Appl Mater Interfaces 9(26):21959–21970

Li X, Gong Y, Zhou X, Jin H, Yan H, Wang S, Liu J (2016) Facile synthesis of soybean phospholipid-encapsulated MoS_2 nanosheets for efficient in vitro and in vivo photothermal regression of breast tumor. Int J Nanomedicine 11:1819–1833

Li J, Wang Y, Zhang L, Xu Z, Dai H, Wu W (2019) Nanocellulose/gelatin composite cryogels for controlled drug release. ACS Sustain Chem Eng 7(6):6381–6389

Licciardi M, Volsi AL, Sardo C, Mauro N, Cavallaro G, Giammona G (2015) Inulin-ethylenediamine coated SPIONs magnetoplexes: a promising tool for improving siRNA delivery. Pharm Res 32(11):3674–3687

Licciardi M, Li Volsi A, Mauro N, Scialabba C, Cavallaro G, Giammona G (2016) Preparation and characterization of inulin coated gold nanoparticles for selective delivery of doxorubicin to breast cancer cells. J Nanomater 2078315:9–21

Lin J, Li C, Zhao Y, Hu J, Zhang LM (2012) Co-electrospun nanofibrous membranes of collagen and zein for wound healing. ACS Appl Mater Interfaces 4(2):1050–1057

Lin N, Geze A, Wouessidjewe D, Huang J, Dufresne A (2016) Biocompatible double-membrane hydrogels from cationic cellulose nanocrystals and anionic alginate as complexing drugs codelivery. ACS Appl Mater Interfaces 8(11):6880–6889

Liu J, Chinga Carrasco G, Cheng F, Xu W, Willfor S, Syverud K, Xu C (2016) Hemicellulose-reinforced nanocellulose hydrogels for wound healing application. Cellulose 23(5):3129–3143

Lopez BSG, Yamamoto M, Utsumi K, Aratsu C, Sakagami H (2009) A clinical pilot study of lignin-ascorbic acid combination treatment of herpes simplex virus. In Vivo 23(6):1011–1016

Lucio D, Martínez Oharriz MC, Jaras G, Aranaz P, Gonzalez-Navarro CJ, Radulescu A, Irache JM (2017) Optimization and evaluation of zein nanoparticles to improve the oral delivery of glibenclamide In vivo study using C elegans. Eur J Pharm Biopharm 121:104–112

Luo Y, Teng Z, Wang Q (2012) Development of zein nanoparticles coated with carboxymethyl chitosan for encapsulation and controlled release of vitamin D3. J Agric Food Chem 60 (3):836–843

Mandracchia D, Tripodo G, Trapani A, Ruggieri S, Annese T, Chlapanidas T, Trapani G, Ribatti D (2016) Inulin based micelles loaded with curcumin or celecoxib with effective anti-angiogenic activity. Eur J Pharm Sci 93:141–146

Mansur AA, De Carvalho FG, Mansur RL, Carvalho SM, De Oliveira LC, Mansur HS (2017) Carboxymethylcellulose/ZnCdS fluorescent quantum dot nanoconjugates for cancer cell bioimaging. Int J Biol Macromol 96:675–686

Markstedt K, Mantas A, Tournier I, Martinez Avila H, Haagg D, Gatenholm P (2015) 3D bioprinting human chondrocytes with nanocellulose–alginate bioink for cartilage tissue engineering applications. Biomacromolecules 16(5):1489–1496

Markstedt K, Escalante A, Toriz G, Gatenholm P (2017) Biomimetic inks based on cellulose nanofibrils and cross-linkable xylans for 3D printing. ACS Appl Mater Interfaces 9 (46):40878–40886

Mertaniemi H, Escobedo-Lucea C, Sanz-Garcia A, Gandia C, Makitie A, Partanen J, Ikkala O, Yliperttula M (2016) Human stem cell decorated nanocellulose threads for biomedical applications. Biomaterials 82:208–220

Mikani M, Torabizadeh H, Rahmanian R (2018) Magnetic soy protein isolate–bovine serum albumin nanoparticles preparation as a carrier for inulinase immobilisation. IET Nanobiotechnol 12(5):633–639

Mirza S, Zia I, Jolly R, Kazmi S, Owais M, Shakir M (2018) Synergistic combination of natural bioadhesive bael fruit gum and chitosan/nano-hydroxyapatite: A ternary bioactive nanohybrid for bone tissue engineering. Int J Biol Macromol 119:215–224

Mitchell CM, Davy BM, Halliday TM, Hulver MW, Neilson AP, Ponder MA, Davy KP (2015) The effect of prebiotic supplementation with inulin on cardiometabolic health: rationale, design, and methods of a controlled feeding efficacy trial in adults at risk of type 2 diabetes. Contemp Clin Trials 45(Part B):328–337

Naguib GH, Hosny KM, Hassan AH, Al Hazmi F, Al Dharrab A, Alkhalidi HM, Hamed MT, Alnowaiser AM, Pashley DH (2018) Zein based magnesium oxide nanoparticles: assessment of antimicrobial activity for dental implications. Pak J Pharm Sci 31(1):245–250

Najafi SHM, Baghaie M, Ashori A (2016) Preparation and characterization of acetylated starch nanoparticles as drug carrier: Ciprofloxacin as a model. Int J Biol Macromol 87:48–54

Nguyen D, Hagg DA, Forsman A, Ekholm J, Nimkingratana P, Brantsing C, Kalogeropoulos T, Zaunz S, Concaro S, Brittberg M, Lindahl A (2017) Cartilage tissue engineering by the 3D bioprinting of iPS cells in a nanocellulose/alginate bioink. Sci Rep 7(1):658

Nirmala R, Woo-il B, Navamathavan R, Kalpana D, Lee YS, Kim HY (2013) Influence of antimicrobial additives on the formation of rosin nanofibers via electrospinning. Colloids Surf B Biointerfaces 104:262–267

Ntoutoume GMN, Granet R, Mbakidi JP, Bregier F, Leger DY, Fidanzi-Dugas C, Lequart V, Joly N, Liagre B, Chaleix V, Sol V (2016) Development of curcumin–cyclodextrin/cellulose nanocrystals complexes: new anticancer drug delivery systems. Bioorg Med Chem 26 (3):941–945

Pallavicini P, Arciola CR, Bertoglio F, Curtosi S, Dacarro G, Agostino A, Ferrari F, Merli D, Milanese C, Rossi S, Taglietti A (2017) Silver nanoparticles synthesized and coated with pectin: An ideal compromise for anti-bacterial and anti-biofilm action combined with wound-healing properties. J Colloid Interface Sci 498:271–281

Park JY, Lee YK, Lee DS, Yoo JE, Shin MS, Yamabe N, Kim SN, Lee S, Kim KH, Lee HJ, Roh SS (2017) Abietic acid isolated from pine resin (Resina Pini) enhances angiogenesis in HUVECs and accelerates cutaneous wound healing in mice. J Ethnopharmacol 203:279–287

Patil SN, Paradeshi JS, Chaudhari PB, Mishra SJ, Chaudhari BL (2016) Bio-therapeutic potential and cytotoxicity assessment of pectin-mediated synthesized nanostructured cerium oxide. Appl Biochem Biotechnol 180(4):638–654

Penalva R, Esparza I, Larraneta E, Gonzalez Navarro CJ, Gamazo C, Irache JM (2015) Zein-based nanoparticles improve the oral bioavailability of resveratrol and its anti-inflammatory effects in a mouse model of endotoxic shock. J Agr Food Chem 63(23):5603–5611

Pereira RF, Sousa A, Barrias CC, Bartolo PJ, Granja PL (2018) A single-component hydrogel bioink for bioprinting of bioengineered 3D constructs for dermal tissue engineering. Mater Horiz 5(6):1100–1111

Posadowska U, Brzychczy Wloch M, Pamula E (2016) Injectable gellan gum-based nanoparticles-loaded system for the local delivery of vancomycin in osteomyelitis treatment. J Mater Sci: Mater 27(1):9–18

Raguvaran R, Manuja BK, Chopra M, Thakur R, Anand T, Kalia A, Manuja A (2017) Sodium alginate and gum acacia hydrogels of ZnO nanoparticles show wound healing effect on fibroblast cells. Int J Biol Macromol 96:185–191

Ranjbar Mohammadi M, Rabbani S, Bahrami SH, Joghataei MT, Moayer F (2016) Antibacterial performance and in vivo diabetic wound healing of curcumin loaded gum tragacanth/poly (ε-caprolactone) electrospun nanofibers. Mater Sci Eng C 69:1183–1191

Rashad A, Mohamed Ahmed S, Ojansivu M, Berstad K, Yassin MA, Kivijaarvi T, Heggset EB, Syverud K, Mustafa K (2018) Coating 3D printed polycaprolactone scaffolds with nanocellulose promotes growth and differentiation of mesenchymal stem cells. Biomacromolecules 19(11):4307–4319

Reesi F, Minaiyan M, Taheri A (2018) A novel lignin-based nanofibrous dressing containing arginine for wound-healing applications. J Drug Deliv Sci Technol 8(1):111–122

Richter AP, Brown JS, Bharti B, Wang A, Gangwal S, Houck K, Hubal EAC, Paunov VN, Stoyanov SD, Velev OD (2015) An environmentally benign antimicrobial nanoparticle based on a silver-infused lignin core. Nat Nanotechnol 10(9):817–823

Rodriguez Del Rio E, Marradi M, Calderon Gonzalez R, Frande Cabanes E, Penades S, Petrovsky N, Alvarez-Dominguez C (2015) A gold glyco-nanoparticle carrying a listeriolysin O peptide and formulated with Advax™ delta inulin adjuvant induces robust T-cell protection against listeria infection. Vaccine 33(12):1465–1473

Salami MA, Kaveian F, Rafienia M, Saber-Samandari S, Khandan A, Naeimi M (2017) Electrospun polycaprolactone/lignin-based nanocomposite as a novel tissue scaffold for biomedical applications. JMSS 7(4):228–238

Sengupta S, Bhattacharyya DK, Goswami R, Bhowal J (2019) Emulsions stabilized by soy protein nanoparticles as potential functional non-dairy yogurts. J Sci Food Agric 99(13):5808–5818

Sharma G, Pathania D, Naushad M (2014) Preparation, characterization and antimicrobial activity of biopolymer based nanocomposite ion exchanger pectin zirconium (IV) selenotungstophosphate: application for removal of toxic metals. J Ind Eng Chem 20 (6):4482–4490

Siddiqui L, Mishra H, Mishra PK, Iqbal Z, Talegaonkar S (2018) Novel 4-in-1 strategy to combat colon cancer, drug resistance and cancer relapse utilizing functionalized bioinspiring lignin nanoparticle. Med Hypotheses 121:10–14

Singla R, Soni S, Padwad YS, Acharya A, Yadav SK (2017) Sustained delivery of BSA/HSA from biocompatible plant cellulose nanocrystals for in vitro cholesterol release from endothelial cells. Int J Biol Macromol 104:748–757

Singla R, Abidi SMS, Dar AI, Acharya A (2019a) Inhibition of glycation-induced aggregation of human serum albumin by organic–inorganic hybrid nanocomposites of Iron oxide-functionalized Nanocellulose. ACS Omega 4(12):14805–14819

Singla R, Abidi SMS, Dar AI, Acharya A (2019b) Nanomaterials as potential and versatile platform for next generation tissue engineering applications. J Biomed Mater Res B 019 (107B):2433–2449

Sun F, Nordli HR, Pukstad B, Gamstedt EK, Chinga-Carrasco G (2017) Mechanical characteristics of nanocellulose-PEG bionanocomposite wound dressings in wet conditions. Mech Behav Biomed 69:377–384

Sun Q, Luan L, Arif M, Li J, Dong QJ, Gao Y, Chi Z, Liu CG (2018) Redox-sensitive nanoparticles based on 4-aminothiophenol-carboxymethyl inulin conjugate for budesonide delivery in inflammatory bowel diseases. Carbohydr Polym 189:352–359

Szu SC, Lin KFY, Hunt S, Chu C, Thinh ND (2014) Phase I clinical trial of O-acetylated pectin conjugate, a plant polysaccharide based typhoid vaccine. Vaccine 32(22):2618–2622

Tang J, Song Y, Tanvir S, Anderson WA, Berry RM, Tam KC (2015) Polyrhodanine coated cellulose nanocrystals: a sustainable antimicrobial agent. ACS Sustain Chem Eng 3 (8):1801–1809

Tansaz S, Boccaccini AR (2016) Biomedical applications of soy protein: a brief overview. J Biomed Mater Res A 104(2):553–569

Tavakolian M, Okshevsky M, Van De Ven TG, Tufenkji N (2018) Developing antibacterial nanocrystalline cellulose using natural antibacterial agents. ACS Appl Mater Interfaces 10 (40):33827–33838

Teng Z, Luo Y, Wang Q (2012) Nanoparticles synthesized from soy protein: preparation, characterization, and application for nutraceutical encapsulation. J Agric Food Chem 60 (10):2712–2720

Tripodo G, Pasut G, Trapani A, Mero A, Lasorsa FM, Chlapanidas T, Trapani G, Mandracchia D (2015) Inulin-D-α-tocopherol succinate (INVITE) nanomicelles as a platform for effective intravenous administration of curcumin. Biomacromolecules 16(2):550–557

Vahedikia N, Garavand F, Tajeddin B, Cacciotti I, Jafari SM, Omidi T, Zahedi Z (2019) Biodegradable zein film composites reinforced with chitosan nanoparticles and cinnamon essential oil: physical, mechanical, structural and antimicrobial attributes. Colloids Surf B: Biointerfaces 177:25–32

Visan AI, Popescu-Pelin G, Gherasim O, Grumezescu V, Socol M, Zgura I, Cristescu R (2019) Laser processed antimicrobial nanocomposite based on polyaniline grafted lignin loaded with gentamicin-functionalized magnetite on a silver-infused lignin core. Polymers 11(2):283–305

Volsi AL, De Aberasturi DJ, Henriksen-Lacey M, Giammona G, Licciardi M, Liz-Marzan LM (2016) Inulin coated plasmonic gold nanoparticles as a tumor-selective tool for cancer therapy. J Mater Chem B 4(6):1150–1155

Wang J, Tian L, Luo B, Ramakrishna S, Kai D, Loh XJ, Meo X (2018) Engineering PCL/lignin nanofibers as an antioxidant scaffold for the growth of neuron and Schwann cell. Colloids Surf B Biointerfaces 169:356–365

Weeks BS, Perez PP (2009) The hemicellulose preparation, Natramune (PDS-2865), increases macrophage phagocytosis and nitric oxide production and increases circulating human lymphocytes levels. Med Sci Monit Basic Res 15(2):43–46

Xue Y, Liang W, Li Y, Wu Y, Peng X, Qiu X, Liu J, Sun R (2016) Fluorescent pH-sensing probe based on biorefinery wood lignosulfonate and its application in human cancer cell bioimaging. J Agric Food Chem 64(51):9592–9600

Xue Y, Mou Z, Xiao H (2017) Nanocellulose as a sustainable biomass material: structure, properties, present status and future prospects in biomedical applications. Nanoscale 9 (39):14758–14781

Yanamala N, Farcas MT, Hatfield MK, Kisin ER, Kagan VE, Geraci CL, Shvedova AA (2014) In vivo evaluation of the pulmonary toxicity of cellulose nanocrystals: a renewable and sustainable nanomaterial of the future. ACS Sustain Chem Eng 2(7):1691–1698

Yao W, Zha Q, Cheng X, Wang X, Wang J, Tang R (2017) Folic acid-conjugated soybean protein-based nanoparticles mediate efficient antitumor ability in vitro. J Biomater Appl 31(6):832–843

Yin K, Divakar P, Wegst UG (2019) Plant-derived Nanocellulose as structural and mechanical reinforcement of freeze-cast chitosan scaffolds for biomedical applications. Biomacromolecules 20(10):3733–3745

Zhang Y, Chen T, Yuan P, Tian R, Hu W, Tang Y, Jia Y, Zhang L (2015) Encapsulation of honokiol into self-assembled pectin nanoparticles for drug delivery to HepG2 cells. Carbohydr Polym 133:31–38

Zhang Y, Jiang M, Zhang Y, Cao Q, Wang X, Han Y, Sun G, Li Y, Zhou J (2016a) Epichlorohydrin-cross-linked hydroxyethyl cellulose/soy protein isolate composite films as biocompatible and biodegradable implants for tissue engineering. ACS Appl Mater Interfaces 8(4):2781–2795

Zhang M, Li X, Li S, Liu Y, Hao L (2016b) Electrospun poly (l-lactide)/zein nanofiber mats loaded with Rana chensinensis skin peptides for wound dressing. J Mater Sci: Mater 27(9):136–148

Zhang Y, Jiang M, Zhang Y, Cao Q, Wang X, Han Y, Sun G, Li Y, Zhou J (2019) Novel lignin–chitosan–PVA composite hydrogel for wound dressing. Mat Sci Eng C-Mater 104:110002–110012

Zhao XJ, Zhang WL, Zhou ZQ (2014) Sodium hydroxide-mediated hydrogel of citrus pectin for preparation of fluorescent carbon dots for bioimaging. Colloids Surf B Biointerfaces 123:493–497

Zhao W, Odelius K, Edlund U, Zhao C, Albertsson AC (2015) In situ synthesis of magnetic field-responsive hemicellulose hydrogels for drug delivery. Biomacromolecules 16(8):2522–2528

Zhou G, Luo Z, Fu X (2014a) Preparation of starch nanoparticles in a water-in-ionic liquid microemulsion system and their drug loading and releasing properties. J Agric Food Chem 62 (32):8214–8220

Zhou P, Xia Y, Cheng X, Wang P, Xie Y, Xu S (2014b) Enhanced bone tissue regeneration by antibacterial and osteoinductive silica-HACC-zein composite scaffolds loaded with rhBMP-2. Biomaterials 35(38):10033–10045

Nanomaterials at the Biological Interphase: Protein Corona Formation and Infusion Reactions

8

Aqib Iqbal Dar, Syed M. S. Abidi, and Amitabha Acharya

Contents

Abstract

The interactions between nanoparticles and macromolecules forming protein corona (PC) introduce new applications and challenges to the nanoparticles (NPs), altering the efficacy and biocompatibility of nanotechnology-based formulations. Accordingly, the overall properties and identity of the biomolecules

A. I. Dar · S. M. S. Abidi · A. Acharya (✉)
Biotechnology Division, CSIR-Institute of Himalayan Bioresource Technology, Palampur, Himachal Pradesh, India

Academy of Scientific & Innovative Research (AcSIR), CSIR-Institute of Himalayan Bioresource Technology, Palampur, Himachal Pradesh, India
e-mail: amitabha@ihbt.res.in

A. Acharya (ed.), *Nanomaterial - Based Biomedical Applications in Molecular Imaging, Diagnostics and Therapy*, https://doi.org/10.1007/978-981-15-4280-0_8

change at this nano-bio interface. Although the nano-bio interface poses hurdles for the effective clinical translation of nanomedicines, it also creates a future window for advancements in nanomedicine. Therefore, this chapter will cover a comprehensive outlook of the advantages and disadvantages of the nano-bio interface with respect to the translational nanomedicine. The implications of nano-bio interface on toxicity, immune recognition, targeting, biodistribution, intracellular uptake, and drug release will also be reviewed. An overview of some of the currently performed research on NP-based clinical translation is also addressed. However, in spite of huge efforts of researchers and other agencies, only few of the nanomedicines made go through the clinical trials. To fill the gap between efforts of the researchers and the clinical translation of nanomedicine, it is need of the hour to collect data from in vivo and in vitro experiments for debugging the nano-bio interfaces. A brief overview of infusion reactions (IRs) outlining different terminologies used to define these reactions and highlighting the gaps in present research nanomedicine-based infusion reactions has also been discussed. Finally, we have presented the strategies to improve the current understanding of these IRs and also to overcome these adverse reactions.

Keywords

Nano-bio interface · Nanoparticles · Protein corona · Protein-nanoparticle interaction · Infusion reactions

Abbreviations

Apo	Apolipoprotein
AuNPs	Gold nanoparticles
Ce6	Chlorin e6
CMF	Chitosan-modified magneto-fluorescent
CNS	Central nervous system
DOPG	1,2-Dioleoyl-sn-glycero-3-phospho-(10-rac-glycerol)
Dox	Doxorubicin
ECM	Extracellular matrix
FA	Folic acid
GO	Graphene oxide
HBO	Hyperbaric oxygen
HC	Hard corona
Ig	Immunoglobulin
MF	Magneto-fluorescent
MTX	Methotrexate
PDT	Photodynamic therapy
PEG	Poly ethylene glycol
RMSNs	RBC-coated mesoporous silica nanoparticles
SPIONs	Superparamagnetic iron oxide nanoparticles
UCNPs	Upconversion nanoparticles
UNPSs	Upconversion nanophotosensitizers

8.1 Introduction

Since the advent of nanotechnology in the scientific world, it has enthralled and got a lot of attention in almost every field of advanced science, viz. physics, chemistry, biology, electronics, nanomedicine, space, communication, etc. The NP-biomolecule interfaces have remarkable potential for both in vitro and in vivo applications, extensively used for targeted drug and gene delivery, biological imaging, tissue engineering, and as fluorescent biological labels for detection of pathogens, for protein identification and sensing, etc. NPs are promising therapeutic delivery vehicles which can host a wide range of active components, including chemotherapeutics, contrast agents, proteins, and nucleic acids, for various biomedical applications. The combination of nanotechnology and biology has resulted in a rapidly advancing field. Since its inception about two decades ago, innovative approaches for using NPs to kill tumours and enhance drug delivery and drug release have been reported in detail. Nanomaterials (NMs), because of their size and ability to interact with other biomolecules, can be considered as "pseudo" subcellular entities that are similar to endogenous biomolecules. The interaction between NPs and biomolecules gives rise to the "nano-bio" interface between the NP and its biological milieu. This can also be defined in terms of the biomolecules that are present on the surface of the NPs.

At the interfaces between the NMs and biological systems, the organic and synthetic worlds merge into a new science, dealing with the safe and effective use of nanotechnology in designing the NMs for various biological applications. The nano-bio interface consists of intricate and dynamic physico-chemical interactions and the kinetic and thermodynamic exchanges between NM surfaces and the biological systems, viz. proteins, lipids, nucleic acids, sugars, small molecule, membranes, organelles, biological fluids, etc. To have a comprehensive understanding of these complex processes, we must understand the different types of dynamic forces and factors that are responsible for shaping these interactions. It is indeed impossible to properly understand and describe all these biophysico-chemical interactions with certainty at these nano-bio interfaces (Kah et al. 2013; Nel et al. 2009).

Over the past few years, there has been an increasing focus on evaluating the important biophysical and physico-chemical identities of protein coronas (PCs) in in vitro and in vivo environments. This has been possible only by using a range of simple as well as sophisticated techniques/methodologies, derived from the vast knowledge of different aspects of science like biophysics, biochemistry, toxicology, colloidal science, molecular biology, etc. (Ke et al. 2017). The most crucial efforts come from the field of nanotechnology and nanomedicine, where nanomaterials (NMs) are synthesized and designed for applications, viz. disease diagnostics, targeting, and drug delivery.

Almost 12 years back, the idea of PC was given by Dawson, Linse, and co-workers (Cedervall et al. 2007). Although it was previously known that the adsorption of proteins occurs onto the surfaces, including particles, its influence on the nano-bio interface was not fully brought into consideration prior to this report. Over more than a decade past, we have learned that the PC formation occurs because of its thermodynamic equilibrium in an aqueous environment, especially minimizing the free enthalpy by the occurrence of some important interaction force, viz. Coulomb's electrostatic interactions, weak van der Waals forces, hydrogen bonding, and hydrophobic interactions. These different interactions at the nano-bio interface ultimately and inevitably give rise to the formation of PC around the NP surface. PC could be generally represented as a double-layer coating adsorbed or wrapped on the NP, consisting of a first tightly bound protein monolayer, usually referred as "hard" corona, composed of proteins having high binding affinity for the NP surface, whereas the outer layer, termed "soft" corona, consisting of a loosely bound protein layer because of its low binding affinity (Lundqvist et al. 2008; Cedervall et al. 2007; Walczyk et al. 2010; Lynch et al. 2009).

In recent years, engineering and designing of NPs and its applicability have expanded its horizons in many scientific branches of science, viz. biology, physics, chemistry, electronics, computer science, engineering, etc. (Atsumi and Belcher 2018; Cox et al. 2018). NPs have many exceptional chemical, optical, mechanical, electrical, and magnetic properties which are quite different from their bulk counterparts (Talapin and Shevchenko 2016; Wetterskog et al. 2013). NPs are promising diagnostic and therapeutic delivery vehicles which can host a wide range of active components, including chemotherapeutics, contrast agents, proteins and nucleic acids, etc., for various biomedical applications (Jin et al. 2018; Zeng et al. 2018). Now it is well known that whenever we use any type of nanosystems for the biological application purpose, the NPs inevitably form biomolecular corona on interacting with different biomolecules present in the interstitial body fluids. Upon the formation of PC, the biological identity of the NP changes significantly. These imparted changes are quite different from the pristine NPs (Monopoli et al. 2012; Lundqvist et al. 2011; Tenzer et al. 2013; Digiacomo et al. 2017). Thus, PC formation can have critical effects on the NPs' biophysico-chemical properties as well as reactivity (Saptarshi et al. 2013). Because of the critical importance of understanding protein-NP interactions and NPs fate inside the living system (Milani et al. 2012; Kim et al. 2016), it is indeed vital to understand the composition, dynamics, and also the kinetics of PC for developing the next-generation nanomedicines for biomedical applications.

Although several aspects of NP-protein interactions, such as physico-chemical and biophysical studies of the PC, designing of stealth NPs, complement activation, current knowledge of NP-proteins interactions with respect to the blood proteins and their toxicity concern, and also the NP elimination from the living system, are discussed thoroughly in recent reviews by a number of research groups, viz. Parak, Caruso, Mahmoudi, or Caracciolo, more extensive and comprehensive studies with deep knowledge are required for proper understanding of biophysico-chemical identity of the PC (Ke et al. 2017; Rahman et al. 2013).

Infusion-related reactions or infusion reactions (IRs) can be defined as the symptoms observed by the patients after the administration of any kind of pharmaceutical product, whether simple or nanoformulation based. There is a lot of diversity in the clinical manifestations/symptoms of these reactions. The most severe forms of IRs may even result in patient death. The symptoms of these reactions may occur within hours to days depending on various biophysico-chemical parameters of the individual as well as the pharmaceutical product. Generally, in more severe cases, the onset of IRs is very rapid (Scarlet 2006). Thus in order to manage these severe IRs caused by the many nanoformulation-based products, there is a prime need to improve the present knowledge based on the clinical trials of these nanomedicines which will help researchers to remove the barriers of complexity of these IRs. For managing these IRs, it is indeed important that risks and consequences are understood as early as possible and also to a bigger picture, in developing strategies and roadmap for overcoming infusion reactions to nanomedicines.

8.2 The Nano-Bio Interface

On entering, into the circulatory system of the human body, NPs circulate throughout the system, and these NPs are then exposed to a complex biochemical milieu, such as blood, lymph, and many other small and large biomolecules. Most abundantly, in the blood, NPs encounter the proteins like serum albumin, apolipoproteins, and also many immunoglobulins and complement proteins (Dar et al. 2019; Nierenberg et al. 2018). As the time passes, these interactions between many of the resident biomolecules inevitably bind or coat to these NPs, ultimately forming a PC. The term protein "coronas" was coined by Cedervall et al., in 2007, describing these as the complex protein structures, having a considerable thickness of 20 to 30 nm. These PCs consisting of soft and hard layers of different kinds of proteins are referred as the soft and hard corona, respectively, based on their binding affinities and efficiencies with the NPs. The soft corona (SC), as the name indicates, is composed of low affinity proteins involving a temporary interaction with the NP (Cedervall et al. 2007). In comparison to this, the hard corona (HC) is formed due to strong interactions of proteins or other biomolecules with the NPs. These interactions arise due to higher affinity of these proteins towards the NPs; hence these interactions are permanent and are stable under the in vitro or in vivo environmental conditions. It is now well known that, mostly, the proteins that form the hard corona are the ones that initially interact with the NPs. After the formation of HC, the soft corona-forming proteins bind to the hard corona by different interaction forces rather than the NP core because all the surface of the NP is covered by the HC proteins (Walczyk et al. 2010). More than a decade ago, Cedervall and co-workers reported that the hard corona proteins bind to the NP for several hours contrary to the soft corona proteins which have a very short lifespan of few minutes for binding to the NP. The strong binding of the HC is generally related to their small molecular sizes and weights compared to the SC proteins (Cedervall et al. 2007). As the PC is formed on the surface of NP, the previously available functional moieties that were solely responsible for different biological responses will now be masked by PC giving a new biological identity and probably the function too. Walkey and Chan proposed that the proteins forming the HC are more crucial in defining the biological

Fig. 8.1 Formation of hard and soft corona in around the NP upon its dynamic interactions with biological fluids, resulting in the formation of hard and soft corona, as shown by green, blue, and cyan globules. (Adapted from Ke et al. 2017, reproduced with permission)

response of the NMs than SC proteins (Walkey and Chan 2012). Thus SC-HC interface is very important for establishing the NP biological identity. The importance of the HC comes from the fact that the isolation of HC is easier than the SC because the SC is transient and challenging in isolation, and thus most of the research has been done or going in the present is related to the HC (Fig. 8.1).

The formation of PC is determined by a number of factors related to the NP and environmental conditions in which they are present. The NP factors include the size, shape, charge, and type/surface of material, whereas the environmental factors comprise of exposure time, composition, pH, temperature, and shearing stress (Rahman et al. 2013). The formation of PC has a huge impact on changing the size, surface charge, surface chemistry, solubility, and aggregation of the NP; hence it can possibly affect the cellular uptake, biodistribution, and macrophage clearance of the NPs under consideration. Thus, controlling the interaction between the NPs and the biomolecules at the bio-nano interface represents a big challenge in the field of nanomedicine. Uncontrolled NP-protein interactions can have a lot of disastrous effects on the biological system. It can mark an NP for its enhanced uptake by macrophage and other cells, increased biodistribution and phagocytosis, and modified signalling pathways and activate enzymatic cascades, or it can also prevent the efficient removal of toxic NPs from the body (Aggarwal et al. 2009; Dutta et al. 2007). An NP can be considered safe only when we have a proper understanding of its controlled physiological response and interaction with its immediate environment. Thus, for the development of safe and effective nanomedicines, it is indeed very crucial and important to have a proper understanding of NPs' biological identity and its physiological response.

8.3 Changes Occurring to the Biomolecules at the Nano-Bio Interface

During the interaction of proteins with the NPs, there are chances of change in the secondary as well as tertiary structural conformations of the proteins, which may be reversible or irreversible (Dar et al. 2019; Fleischer and Payne 2014; Liu et al. 2013). A particular correlation between the size of NP and protein unfolding has been observed, the larger NPs having a lower surface area because of more changes in the protein structural conformation (Teichroeb et al. 2008; Lundqvist et al. 2004). Since the function of proteins is mainly dependent upon the structure of the protein, the formation of PC may alter the biological identity and function of these biomolecules. Alternatively, when the proteins come in close contact to each other at the surface of the NP, there are chances of aggregation of these proteins. These aggregations can lead to the protein fibril formation, which ultimately can be a potential reason for causing a number of neurodegenerative diseases like Parkinson's, Huntington's, and Alzheimer's diseases and type II diabetes mellitus (Singla et al. 2019; Espay et al. 2019; Ma et al. 2019).

8.4 Changes Occurring to the Nanoparticles at the Nano-Bio Interface

After the NP-protein interaction, the physico-chemical characteristics of the NPs are likely to change considerably in the biological milieu. These properties include the shape, size, charge, and surface of the NP under consideration. In general, the size of the NPs increases due to the formation of PC. This increase in the size depends upon the type of NP, its size, shape, charge, and protein source. The formation of PC generally occurs in multilayers; however, the factors that determine the number of layers remain elusive and unanswered. There are also chances of decrease in the size of NPs (certain lipid NPs) in the presence of protein milieu, apparently due to the osmotic forces that drag these NPs to shrink (Wolfram et al. 2013). Besides this, the PC may trigger nanoparticle aggregation and fibrillation through different forces, resulting in the formation of larger aggregates of proteins in both in vitro and in vivo systems (Singla et al. 2019; Tanokashira et al. 2017; Konar et al. 2019). In contrast to this, the formation of PC around the NP can also stabilize and prevent these particles from aggregation (Albanese and Chan 2011; Wells et al. 2012).

8.5 Consequences and Implications of the Nano-Bio Interface

As the physico-chemical properties of both the NPs and biomolecules change at the nano-bio interface, such changes may affect the biocompatibility/efficacy of the nanomedicine. Hence, for the development of more effective nanoformulations for the clinical translation of the nanotechnology in the field of nanomedicine, the understanding of PC at the nano-bio interface becomes a necessity. In this section we will be discussing about certain important implications of protein-NP interactions that must be understood before using any type of NP in translational nanomedicine (Table 8.1).

8.5.1 Toxicity Concerns

The safety and toxicity evaluation of the nanoformulations used in the field of nanomedicine is of utmost importance so as to provide a safe and effective means of modern medicine. As the properties of NPs are unique than their bulk counterparts (Gatoo et al. 2014; Ansari et al. 2019; Breznan et al. 2018), there should be a comprehensive and thorough evaluation of these nanomedicines at the physiological, cellular, molecular, genetic, and organ level. However, the development of the protocols for the evaluation of potential toxicity of these NPs is a challenging task. This is mainly because of the rapid variation in the physico-chemical properties of NPs upon the interference with the biomolecules. The NP interaction with blood proteins can even cause toxicity like different types of hypersensitivity generally called as infusion reactions (IRs) and also thrombosis in many of the cases (Moghimi 2018; Corbo et al. 2016; Szebeni et al. 2018a). There are many literature reports which suggest that the biocompatibility of NPs can be improved, provided the NP favours the preferential adsorption of albumin proteins on its surface (Corbo et al. 2016; De Paoli et al. 2014). The longer circulation time of the NPs can make them more reliable for their possible therapeutic or diagnostic use, due to prolonged contact with biological milieu in the living system. However, making stealth NPs invisible to the immune system, there are increased possibilities that it may also change the toxicity profile of these NPs. So, in conclusion, we can say that the inevitable formation of the PC is a necessary evil having both positive and negative effects on the living system. The PC-NP complexation can increase the biocompatibility of the NP, also decreasing the cell adhesion and cell free energy. PC at its negative side can decrease the targetability of otherwise targeted NPs; an immediate and intense immune response in the form of infusion reactions can occur. The PC formation can also make a previously biocompatible NP susceptible to enter into the autophagy pathway of the cell, making it toxic for the cellular environment. In other words, PC can be seen to have two-faced (Janus)-like behaviour. Sometimes it directs the NPs off the road but also provides many properties that can make these NPs more effective/specific in their function.

8.5.2 Recognition by the Immune System

The major obstacle for the successful delivery of nanoformulation into the targeted site is the presence of the reticuloendothelial system (RES) all over the individual's body (Bertrand et al. 2017; Oh et al. 2018; Mirshafiee et al. 2016). After entering into the system, these NPs are generally phagocytosed by immune cells present in the RES organs like macrophages, lymph nodes, the bone marrow, the liver, etc. Thus,

Table 8.1 Few literature reports describing the recent advances in PC-based theranostic applications in translational nanomedicine

NMs used	Theranostic application/s	Conclusions	Reference
DOPG-liposome	CNS cancer diagnosis	Using DOPG-liposome-based diagnostic method, HCs of meningeal tumour patients could be discriminated from those of healthy subjects.	Papi et al. (2019a)
		The results suggested that the concept of HC-based diagnostic methods can pave the path for early-stage diagnosis of CNS cancer with high specificity and sensitivity.	
GO flakes	Pancreatic cancer	Early detection of pancreatic cancer using GO flake-based PC with 92% sensitivity.	Papi et al. (2019)
		The technique is very accurate, cost-effective, quick, and easy, and we believe that the PC-based blood test has the potential to become a diagnostic marker in early detection of cancer and other diseases.	
MF and CMF	Immunoglobulin detection	The rapid immunoglobulin diagnosis can be possible using this PC formation concept under in vitro and ex vivo conditions.	Dar et al. (2019)
		We believe that the diseases involving elevated levels of immunoglobulins, viz. rheumatoid arthritis, diabetic nephropathy, etc., can possibly be using this PC formation concept.	
AuNPs	Immunoglobulin detection	Simple and fast blood test for point-of-care clinics or field settings to evaluate humoral immune status of animals.	Zheng et al. (2017)
		The test can also be used for determining the active immune response during the infection.	
PEGylated, peptide-coated SPIONs	Heart diseases (atherosclerosis)	The results suggest that in the future these modified SPIONs may be used in early detection of heart diseases such as atherosclerosis and aortic stenosis.	Meisel et al. (2018).
MTX NP@lipid-FA	Cancer therapy	A superior therapeutic effect of MTX NP@lipid-FA over the free drug was found on tumours under in vivo conditions due to better	Zeng et al. (2018)

(continued)

Table 8.1 (continued)

NMs used	Theranostic application/s	Conclusions	Reference
		profiles in pharmacokinetics and biodistribution.	
		In the future, we expect this concept can become a generalized approach for NP drug production having direct impact on clinical translation.	
$NaGdF_4$:Yb,Er, Ca@$NaYbF_4$: Ca@$NaNdF_4$:Gd,Ca UCNPs	Cancer therapy	The HBO-assisted PDT decomposes collagen present in tumour, facilitating the diffusion of oxygen and thereby increasing the penetration power of these UNPSs.	Li et al. (2018)
		This synergic effect of HBO-assisted and UNPS-mediated PDT results in enhanced therapeutic efficacy at a low laser power as compared with UNPS alone.	
ApoE4-decorated lipid NPs	Brain-related diseases	The PC based on ApoE4-adsorbed surfactant-stabilized NPs allows the specific targeting of the cerebrovascular endothelium.	Dal Magro et al. (2018)
		This strategy increased the translocation of NP into brain parenchyma, improving threefold NP accumulation in the brain as compared to undecorated particles.	
Affibody-functionalized AuNPs [$(Z_{NS1}12)_2$-AuNP]	Dengue	The developed anti-NS1 affibody-functionalized AuNPs [$(Z_{NS1}12)_2$-AuNP] for selective and sensitive detection of dengue NS.	Bang et al. (2018)
		This novel and sensitive method using $(ZNS_1 12)_2$-AuNP may be used in detection of dengue patients at an early stage	
RMSNs-dox/Ce6	Breast cancer	Development of RBC-coated multifunctional MSN-based theranostic NPs with improved colloidal stability, long blood circulation, enhanced drug release, and tumour imaging.	Su et al. (2017)
		The RMSNs could integrate the Ce6 and Dox for simultaneous photodynamic therapy and chemotherapy and can completely suppress breast cancer metastasis.	

(continued)

Table 8.1 (continued)

NMs used	Theranostic application/s	Conclusions	Reference
AuNPs	Alzheimer's disease and multiple sclerosis	Development of AuNP-based diagnostics for the detection of Alzheimer's disease and multiple sclerosis.	Hajipour et al. (2017)
		The colorimetric, chemometric and statistical analysis demonstrated the promising use of this concept for unambiguously identifying and discriminating these two diseases.	
AuNPs	Influenza/viral infections	A novel AuNP-based test for the timely detection and diagnosis of infectious viral diseases.	Zheng et al. (2017)
		This test can be applied for any local, epidemic, or pandemic viral disease-exposed areas for immediate screening of infected individuals.	

major fraction of injected nanoformulations gets accumulated, instead of going to the targeted tissue or cell (Rao et al. 2017; Nguyen and Lee 2017). After this the immunoglobulins can directly bind to the phagocytic cells or indirectly can stimulate the process of phagocytosis through complement activation (Corbo et al. 2016; Ritz et al. 2015). Further, NP-induced changes in the structural conformational of protein can switch on the immune system, as the individual's body is no longer recognizing the proteins as its native counterparts. Hence the overall impact of PC on the immune response can occur in different manners, viz. the PC can induce differential recognition of NPs by immune cells, activation of complement system by PC-NP complexation, and induction of adaptive immune response by PC-NP, and finally it can evade the whole immune system (Yan et al. 2013; Lee et al. 2015; Le et al. 2019; Oh et al. 2018).

8.5.3 Molecular Targeting

The targeted delivery of nanoformulations using NPs can abate its side effects, with enhanced efficacy. However, the translation of these NP-based formulations into practical clinical applications remains a challenge from decades, which is mainly due to upregulating interactions at the nano-bio interface in a living system. Proper understanding of interactions between the cell membrane and the PC surrounding the NPs is crucial for designing the NP-based nanoformulations with a good targeting ability for theranostic applications. Because of the complexity of cell membrane components due to its dynamic interactions with other biomacromolecules present in the environment (Nel et al. 2009; Mahmoudi et al. 2013), there are new opportunities as well as challenges for better drug delivery and targeting, to be used for theranostic applications by engineering the nanosystem. Therefore, for designing of NP-based delivery systems, we must have a deep knowledge of interaction between the cell membrane and the NP corona. Over the past years, a large number of nanoformulations have been developed time to time. However, it is unfortunate that the actual benefits from these targeting molecules have not been so promising as it was expected to be (Al-Ahmady et al. 2018; Quijada et al 2019; Obst et al. 2017; Assali et al. 2019). The most important reason being the modification of biological and chemical identities of the targeted nanosystems. Further, as it is well known that the protein corona has a dual effect, it can act as enhancer for the targeted delivery to the tissues and cells, but at the same instant, it can encourage the NP clearance from the blood due to phagocytosis by the RES system in synchronization with the immune response. Therefore, more detailed and comprehensive studies are still required for the engineering of NP-PCs having optimum and desired properties.

8.5.4 Biodistribution

The NP's fate inside the body is a major concern for the safety with respect to the translational medicine. The removal of many NPs from the system is a question of minutes, but the interaction of these NPs with the cells of other distant organs can be related for hours after exposure depending on the type of NP under consideration. This ultimately results in the failure of that NP as a candidate for clinical application. For curbing these anomalies, the emerging concept is using the PC for conferring the transport properties to the NPs and also providing a better and modified biological identity to them (Monopoli et al. 2012; Caracciolo et al. 2019). In this essence, the biodistribution of NPs is governed by the protein coating, as the proteins exhibit the differential affinities at the tissue level (Choi et al. 2007; Cabral et al. 2011; Treuel et al. 2014; Monopoli et al. 2011; Ritz et al. 2015; Mirshafiee et al. 2016). As it is well known that the PC composition has a significant effect on the NP biodistribution, for example, the adsorption of opsonins (fibrinogen, IgG, complement proteins) on the NPs marks them for phagocytosis and subsequent removal from the system. In contrast, dysopsonin (HSA, apolipoproteins, RBC membrane proteins) adsorption has opposite effect, increasing the circulation time and enhancing transport across the otherwise non-accessible membranes like the blood-brain barrier (Zhang et al. 2019; Rao et al. 2017; Oh et al. 2018; Lynch and Dawson 2008; Kharazian et al. 2018; Giulimondi et al. 2019).

8.5.5 Intracellular Uptake of Nanoparticles

The understanding of protein-NP interactions is very crucial for the development of NPs with targeted delivery. Also, knowing the PC composition, a number of proteins thereof could contribute for controlling the intracellular uptake of these NPs.

Different constituents of the PC on nanocarriers have been determined by a number of studies (Tenzer et al. 2013; Kah et al. 2014), but the complexity creates a challenge to predict its impact. The current gold standard is to optimize the NP properties for the fruitful exploitation of PC for specific biomedical applications. Studies have shown a correlation between NP protein content and degree of cellular uptake. On the other hand, it has also been shown that the PC formation increases the intracellular uptake of the NPs (Rodriguez 2013). Furthermore, the PC formation can also change the intracellular uptake mechanism caused by a size-dependent effect through protein-induced NP aggregation. This gives an illustration about the impact of PC on the NP interactions with the cellular machinery (Pelaz et al. 2015; Treuel et al. 2014; Schöttler et al. 2016; Ritz et al. 2015; Wolfram et al. 2014).

8.5.6 Drug Release

The advent of nanocarrier-based drug delivery system has significantly contributed towards the rapid development of new and improved therapeutic agents against a number of acute diseases. However, for successful translational application in nanomedicine, the nanodelivery system has some limitations. These limitations arise mainly due to the lack of controlled release in a sustained manner. A plethora of modern and sophisticated approaches have been employed for curbing these limitations. But, despite the development of various types of these new nanocarrier systems, the effect of PC formation on the drug release profile is still not comprehensively evaluated. This is because the PC formation on the nanocarrier system changes its biological identity as well as its function property (Al-Ahmady et al. 2018; Behzadi et al. 2014). As the physico-chemical characteristics of NPs like size, shape, charge, and surface morphology change due to PC formation, there are significant chances that the drug release profile may be enhanced or hindered (Quijada et al 2019; Obst et al. 2017; Assali et al. 2019).

8.6 Opportunities and Advancements: A Future Window

Although there are several problems related to the bio-nano interface at the NP delivery systems, a deep knowledge and comprehensive understanding of NP-PC interactions may provide a new path for a better delivery system at translational level. For instance, the NPs' surface binding proteins can give a blueprint of the different biomacromolecules that are adsorbed on them as PC. So, it can provide a protein molecular signature of the source under consideration. Accordingly, a comprehensive proteomic and systematic analysis of the different proteins and PCs needs to be done. The impact of surfactant coatings/modifications could be useful in identifying the NPs that have strong affinity for the proteins of interest. These new strategies will enable us to have a good control over the protein binding. This would also provide numerous opportunities to improve NP biocompatibility, biodistribution, release, and efficacy thereof. Extensive efforts are needed to be made

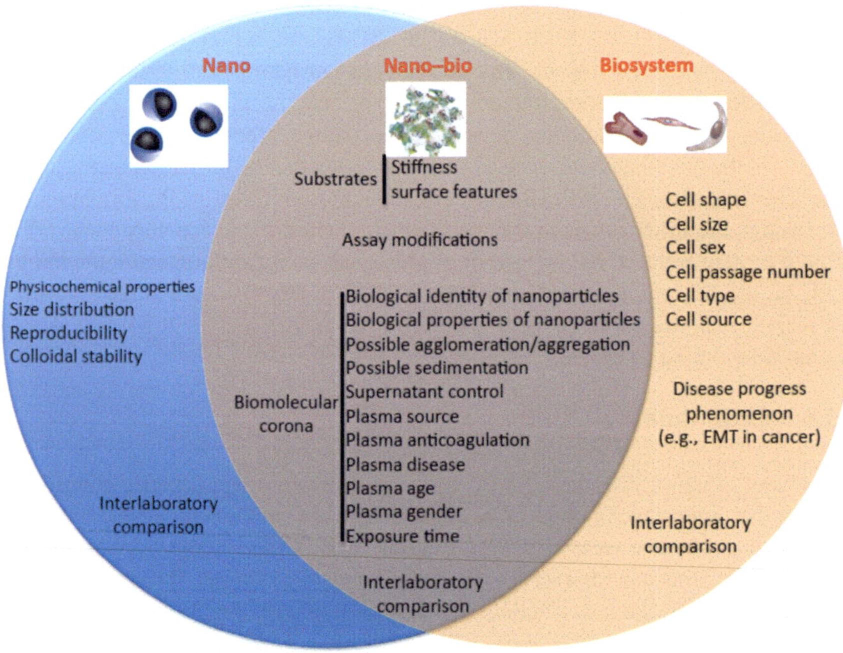

Fig. 8.2 Schematic illustration showing the minimal set of required information and experimental setups required in understanding the nano-bio interface parameters, for achieving robust and reliable nanomedicine readouts. (Adapted from Mahmoudi 2018, reproduced with permission)

in the field of nanomedicine for a proper clinical translation. But there are a number of factors that need to be addressed in the future (Ledford 2016; Panet et al. 2017). A detailed and comprehensive knowledge of the components of nano-bio interface, properties including shape, size, surface morphology, colloidal stability, and protein conformation and biological properties like protein source, cell sex, cell age, cell type, and the experimental setups, are required. All these need to be taken into consideration for filling the gap between the scientific efforts and effective translational nanomedicine (Fig. 8.2). Based on all these parameters that are needed in the clinical translational nanomedicine, the futuristic efforts should rely on collecting all the small and big data information on the protein corona for constructing a robust library of nanomedicine. This will critically help in successful translation of nanotechnologies in nanomedicine (Dar et al. 2019; Mahmoudi 2018).

A detailed account of some of the closely related literature reports describing the recent advances in research for debugging the complex nano-bio interfaces to accelerate the clinical translation in the field of nanomedicine has also been comprised in Table 8.1.

8.7 Infusion Reactions

Infusion reactions (IRs) are the adverse drug reactions with intricate, immune-responsive side effects caused by an intravenously administered pharmaceutical product used for a diagnostic or a therapeutic purpose (Szebeni et al. 2018a; Joerger 2012; Johansson 2001; Jutel 2015; Lenz 2007). These reactions generally occur within minutes to hours after receiving a particular therapeutic dose. There is a lot of diversity of these administered products, viz. the traditional ones include the biological agents (recombinant proteins/antibodies), drugs of low molecular weight, and nucleic acids (Khan 2016; Patel and Khan 2017; Sampson 2006) and the modern type includes the nanomedicine products (Lorenzo-Abalde and Gonzalez-Fernandez 2013; Szebeni 2012a, Szebeni et al. 2018a). Though the IRs are not confined to the nanotechnology-based medicines, they represent a big snag for the translation of nanomedicines. The IRs are sometimes also called as hypersensitivity or anaphylaxis, but the defining terminology changes, depending on the field of study, viz. immunology or clinical oncology. All the IRs caused by these formulations have differences in their mechanisms of action, but despite this difference, almost all the type of IRs generally have the overlapping symptoms and signs. The most common clinical manifestations of the patients with severe IRs include the flushing or rashes in skin, pain in back and chest, wheezing, difficult breathing, fever, chills, rigor, etc. (Table 8.2). These adverse effects caused by the IRs require an accurate and timely evaluation, for its management and also to avoid its potentially severe consequences. There are very less chances ($<5\%$) for the occurrence of a severe type of IRs in patients (Lenz 2007). However, the propensity can increase when a combination of different drug products is used, viz. carboplatin and paclitaxel. Regardless of a lot of research that has been done from decades for elucidating the incidence of IRs, till now it is believed that it depends on the pharmacodynamics and pharmacogenetics of the drug (Markman et al. 2000).

8.7.1 Terms Used to Report IRs

8.7.1.1 Anaphylaxis

According to Coombs and Gell's classification of hypersensitivity reactions (Coombs and Gell 1975; Simons et al. 2015), the anaphylaxis can be considered as severe, life-threatening, generalized, or systemic hypersensitivity reactions. However, the actual definition of anaphylaxis is still evolving and suggests that there is no agreement in regard with the development of anaphylaxis, particularly, whether non-IgE-based mechanisms may or may not be involved.

8.7.1.2 Anaphylactoid Reaction

Anaphylactoid reactions are immediate systemic reactions mimicking the anaphylaxis but are non-IgE-mediated factors from mast cells and basophils (Cox et al. 2007; Sampson et al. 2006). At the first exposure of antigen, the anaphylactoid reactions may be clinically indistinguishable from that of the anaphylaxis. The

Table 8.2 Some examples of IR-inducing nanotechnology-based drug products and their symptoms

Name	Active ingredient	Indication	Particle type	Symptoms
Onivyde	Irinotecan	Pancreatic adenocarcinoma	Liposomes	Rash, urticaria, oedema
Myocet	Doxorubicin	Multiplex	Liposomes	Flushing, dyspnoea, fever, headache, chills, hypotension
Doxil	Doxorubicin	Ovarian cancer, Kaposi sarcoma	Liposomes	Flushing, breathlessness, tightness in the chest/throat, facial swelling, chills, back pain, hypotension
Abelcet	Amphotericin B	Fungal infections	Solid microparticles	Breathlessness, hypo-/hypertension
Ambisome	Amphotericin B	Fungal infections	Liposomes	Chills, rigors, fever, nausea, vomiting, cardiorespiratory events
DaunoXome	Daunorubicin	Kaposi sarcoma	Liposomes	Back pain, flushing, chest tightness
Taxol	Paclitaxel	Cancer	Cremophor EL	Anaphylaxis, angioedema, arrhythmias, chills, dyspnoea, fever, rash, sudden death, tachycardia, urticaria, wheezing
Etoposide	Podophyllotoxin	Cancers	Polysorbate 80	Apnoea, back pain, dyspnoea, fever, flushing chills, coughing, cyanosis, diaphoresis, facial swelling, rash, tachycardia, swelling, urticaria
Cyclosporine injection	Cyclosporine	Immunosuppression	Cremophor EL	Acute respiratory distress, anaphylaxis, angioedema, chills, dyspnoea, flushing, fever, urticaria, wheezing, tachycardia
Fasturec	Rasburicase	Hyperuricaemia	Poloxamer 188	Anaphylaxis, hypotension, nausea bronchospasm, chest pain, rash, rhinitis, urticaria, vomiting, diarrhoea, dyspnoea, fever

anaphylactoid reactions are different from the anaphylaxis as these are milder upon repeated administrations. However, this does not give a significant clue to separate the anaphylactoid reaction from anaphylaxis.

8.7.1.3 Complement Activation-Related Pseudoallergy

Complement activation-related pseudoallergy (CARPA) is a type of hypersensitivity reaction, having similarity in the symptoms of Coombs and Gell's type I category of hypersensitive reactions. These hypersensitive reactions do not involve the pre-existing IgE antibodies but arise due to the activation of complement system (Szebeni 2005). The activation of complement causes the release of other complement proteins (C3a, C5a, and C5b-9), which creates a trigger for basophils, mast cells, and other phagocytic cells, causing the release of vasoactive mediators (Szebeni et al. 2011; Szebeni 2012a). CARPA generally occurs within minutes of the infusion. However, it can also be a late reaction mostly in case of premedicated patients. During CARPA, almost all the organs can get affected. The salient features that distinguish non-IgE CARPA from the classical IgE-mediated hypersensitive reactions are as follows: (i) these reactions arise mostly at first exposure but is milder or absent upon re-exposure; (ii) these reactions are resolved spontaneously; (iii) reactions are mostly tachyphylactic, i.e. rapid and short term; (iv) reaction rate is high (2–10% or more); and (v) these reactions cannot be predicted by standard allergy tests.

8.7.1.4 Cytokine Release Syndrome

The cytokine release syndrome (CRS) is an inflammatory response syndrome that arises due to some infections or diseases. The CRS can also occur by using some monoclonal antibody drugs as well as by employing the adoptive T-cell therapies (Lee et al. 2014; Kroschinsky et al. 2017). Sometimes when the severity is high, the CRS is also called as cytokine storms. The CRS is believed to occur by sustained activation of monocytes, macrophages, B cells, and T cells. During CRS, the levels of TNFα and IFNγ increase within 1 or 2 h of exposure, followed by increases in many cytokines. A number of corticosteroids have been used for the effective reduction of the severity of symptoms caused by cytokine release (Bugelski et al. 2009).

8.8 Strategies for Overcoming IRs to Nanomedicines

The hurdles in the translation due to IRs are not unique to nanoformulation-based drugs. Hence, a way out to properly address the problem is to use all the knowledge and information gained from the clinical use of other drugs. For this, the approach must be systematic. The initial step should be the biomarker identification related to a particular disease or patient followed by a proper establishment of in vivo models for understanding the mechanisms of IRs (Kozma et al. 2015; Morgan and Harris 2015).

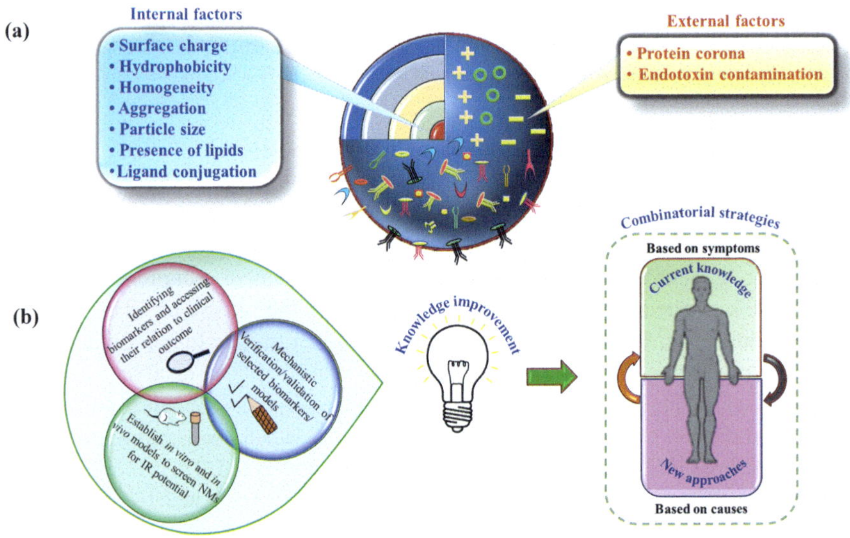

Fig. 8.3 Strategies and roadmap for overcoming infusion reactions to nanomedicines. (**a**) Some physico-chemical attributes of nanomedicines that have been linked to IRs. (**b**) Identification of reliable biomarkers, validation of a methodological framework, and mechanistic verification serve to improve the current knowledge base of IRs to nanomedicines

The current knowledge-based approaches used for slowing down the rate of IRs are powerful tools for reducing IRs to nanomedicines (Szebeni et al. 2011); further it will be a benefit for us to understand the underlying mechanism of these reactions. In the future, the improved knowledge based on the clinical trials of these nanomedicines will allow researchers to unravel the complexity of these IRs, different immune-mediated adverse effects, and also tissue-specific toxicities caused by nanomedicines. The overall strategies to be followed overcoming the nanomedicine-based infusion reactions are described in Fig. 8.3.

8.9 Conclusions and Future Perspectives

With the distinctive properties of nanomedicines, these pose significant advantages over the conventional medicines. However, nanomedicines once administrated into the living system inevitably form PC. These PCs consist of a plethora of complex proteins differing in their relative abundance. This formation of PC on the NP surface ultimately modifies NPs' biophysico-chemical properties giving a new chemical as well as biological identity. To a greater extent, the complexity of PC roots from the complexity of biomacromolecules present in the living system. These structural and functional modifications of NPs cause potentially adverse effects on the living system. These implications may be mild or severe depending upon various NP and biological factors. The important factors and implication that must be addressed when employing any type of NP in translational nanomedicine are toxicity concerns of the nanomedicine, its immunological recognition, targeting ability, biodistribution in the system, its intracellular uptake, and drug release. So, for bridging the gap between the efforts of the researchers and the clinical translation of nanomedicine, it is indeed necessary to collect all the big data from in vivo and in vitro experiments for debugging the nano-bio interfaces. The increased understanding of nano-bio interface will open a new window for the design of modern nanomedicines with improved efficacy, least toxicity, biodistribution, and release into the intracellular environments.

Infusions reactions, as a type of nano-bio interaction, are serious threat caused by the use of different types of conventional as well as nanotechnology-based drug products. There is a lot of diversity in the adversity of these IRs when used in combinations. However, to overcome these adverse effects caused by the IRs, we require an accurate and timely assessment, for its management and also to avoid its potentially severe consequences.

The principal strategies that should be followed to overcome these IRs comprise a comprehensive and deep knowledge emphasizing on the mechanistic understanding of IRs and also to identify the main cause for triggering these IRs. Inclusive knowledge should be gained on the role of different proteins (complement, cytokines) and cells (macrophages, basophils, platelets) in IRs and also their NP-based physico-chemical attributes. This will greatly help in understanding the possible role of long-term and tissue-specific cytotoxicity. Moreover, there is a prime need to standardize specific methodologies, biomarkers, and in vitro/in vivo models for predicting IRs in patients. In the future, all the data and improved knowledge of all these studies must be combined to overcome the IR-related barriers to the clinical translation of nanomedicines.

Acknowledgments The authors would like to thank the director of CSIR-IHBT for his constant support and encouragement. AA acknowledges the financial support from CSIR (MLP-201) and DST (GAP-0214; EMR/2016/003027). AID and SMSA thank DST and DBT for SRF, respectively. The CSIR-IHBT communication number of this manuscript is 4498.

References

Aggarwal P, Hall JB, McLeland CB, Dobrovolskaia MA, McNeil SE (2009) Nanoparticle interaction with plasma proteins as it relates to particle biodistribution, biocompatibility and therapeutic efficacy. Adv Drug Deliv Rev 61(6):428–437

Al-Ahmady ZS, Hadjidemetriou M, Gubbins J, Kostarelos K (2018) Formation of protein corona in vivo affects drug release from temperature-sensitive liposomes. J Control Release 276:157–167

Albanese A, Chan WC (2011) Effect of gold nanoparticle aggregation on cell uptake and toxicity. ACS Nano 5(7):5478–5489

Ansari M, Sinha BB, Phase D, Sen D, Sastry PU, Kolekar YD, Ramana CV (2019) Particle size, morphology, and chemical composition controlled CoFe2O4 nanoparticles with Tunable magnetic properties via oleic acid based solvothermal synthesis for application in electronic devices. ACS Appl Nano Mater 2(4):1828–1843

Assali A, Razzazan S, Akhavan O, Mottaghitalab F, Adeli M, Atyabi F (2019) The bio-interface between functionalized Au NR@GO nanoplatforms with protein corona and their impact on delivery and release system. Colloids Surf B: Biointerfaces 173:891–898

Atsumi H, Belcher AM (2018) DNA Origami and G-Quadruplex hybrid complexes induce size control of single-walled carbon nanotubes via biological activation. ACS Nano 12 (8):7986–7995

Bang J, Park H, Choi WI, Sung D, Lee JH, Lee KY, Kim S (2018) Sensitive detection of dengue virus NS1 by highly stable affibody-functionalized gold nanoparticles. New J Chem 42 (15):12607–12614

Behzadi S, Serpooshan V, Sakhtianchi R, Müller B, Landfester K, Crespy D, Mahmoudi M (2014) Protein corona change the drug release profile of nanocarriers: the "overlooked" factor at the nanobio interface. Colloids Surf B: Biointerfaces 123:143–149

Bertrand N, Grenier P, Mahmoud M, Lima EM, Appel EA, Dormont F, Farokhzad OC (2017) Mechanistic understanding of in vivo protein corona formation on polymeric nanoparticles and impact on pharmacokinetics. Nat Commun 8(1):777–785

Breznan D, Das DD, MacKinnon-Roy C, Bernatchez S, Sayari A, Hill M, Kumarathasan P (2018) Physicochemical properties can be key determinants of Mesoporous silica nanoparticle potency in vitro. ACS Nano 12(12):12062–12079

Bugelski PJ, Achuthanandam R, Capocasale RJ, Treacy G, Bouman-Thio E (2009) Monoclonal antibody-induced cytokine-release syndrome. Expert Rev Clin Immunol 5(5):499–521

Cabral H, Matsumoto Y, Mizuno K, Chen Q, Murakami M, Kimura M, Terada Y, Kano MR, Miyazono K, Uesaka M, Nishiyama N, Kataoka K (2011) Accumulation of sub-100 nm polymeric micelles in poorly permeable tumours depends on size. Nat Nanotechnol 6 (12):815–823

Caracciolo G, Safavi-Sohi R, Malekzadeh R, Poustchi H, Vasighi M, Chiozzi RZ, Di Carlo A (2019) Disease-specific protein corona sensor arrays may have disease detection capacity. Nanoscale Horiz 4:1063–1076

Cedervall T, Lynch I, Lindman S, Berggård T, Thulin E, Nilsson H, Linse S, Dawson KA (2007) Understanding the nanoparticle–protein corona using methods to quantify exchange rates and affinities of proteins for nanoparticles. Proc Natl Acad Sci U S A 104(7):2050–2055

Choi HS, Liu W, Misra P, Tanaka E, Zimmer JP, Itty Ipe B, Bawendi MG, Frangioni JV (2007) Renal clearance of quantum dots. Nat Biotechnol 25(10):1165–1170

Coombs RRA, Gell PGH (1975) Classification of allergic reactions responsible for clinical hypersensitivity and disease. In: PGH G, RRA C, Lachmann PJ (eds) Clinical aspects of immunology, 3rd edn. Blackwell Scientific Publications, Oxford, pp 761–781

Corbo C, Molinaro R, Parodi A, Toledano Furman NE, Salvatore F, Tasciotti E (2016) The impact of nanoparticle protein corona on cytotoxicity, immunotoxicity and target drug delivery. Nanomedicine 11(1):81–100

Cox L, Platts-Mills TA, Finegold I, Schwartz LB, Simons FER, Wallace DV (2007) American academy of allergy, asthma & immunology/American college of allergy, asthma and immunology joint task force report on omalizumab-associated anaphylaxis. J Allergy Clin Immunol 120 (6):1373–1377

Cox A, Andreozzi P, Dal Magro R, Fiordaliso F, Corbelli A, Talamini L, Krol S (2018) Evolution of nanoparticle protein Corona across the blood–brain barrier. ACS Nano 12(7):7292–7300

Dal Magro R, Albertini B, Beretta S, Rigolio R, Donzelli E, Chiorazzi A, Sancini G (2018) Artificial apolipoprotein corona enables nanoparticle brain targeting. Nanomedicine: NBM 14 (2):429–438

Dar AI, Walia S, Acharya A (2019) Molecular recognition based rapid diagnosis of immunoglobulins via proteomic profiling of protein-nanoparticle complexes. Int J Bio Macromol 1(138):156–167

De Paoli SH, Diduch LL, Tegegn TZ (2014) The effect of protein corona composition on the interaction of carbon nanotubes with human blood platelets. Biomaterials 35(24):6182–6194

Digiacomo L, Pozzi D, Amenitsch H, Caracciolo G (2017) Impact of the biomolecular Corona on the structure of Pegylated liposomes biomaterials science. Biomater Sci 5(9):1884–1888

Dutta D, Sundaram SK, Teeguarden JG, Riley BJ, Fifield LS, Jacobs JM, Addleman SR, Kaysen GA, Moudgil BM, Weber TJ (2007) Adsorbed proteins influence the biological activity and molecular targeting of nanomaterials. Toxicol Sci 100(1):303–315

Espay AJ, Vizcarra JA, Marsili L, Lang AE, Simon DK, Merola A, Josephs KA, Fasano F, Savica R, Greenamyre JT, Cambi F, Yamasaki TR, Tanner CM, Gan-Or Z, Litvan I, Mata IF, Zabetian CP, Brundin P, Fernandez HH, Standaert DG, Kauffman MA, Schwarzschild MA, Sardi SP, Sherer T, Perry G, Leverenz JB (2019) Revisiting protein aggregation as pathogenic in sporadic Parkinson and Alzheimer diseases. Neurology 92(7):329–337

Fleischer CC, Payne CK (2014) Secondary structure of corona proteins determines the cell surface receptors used by nanoparticles. J Phys Chem B 118(49):14017–14026

Gatoo MA, Naseem S, Arfat MY, Mahmood Da A, Qasim K, Zubair S (2014) Physicochemical properties of nanomaterials: implication in associated toxic manifestations. Biomed Res Int 2014:1–8

Giulimondi F, Digiacomo L, Pozzi D, Palchetti S, Vulpis E, Capriotti AL, Mahmoudi M (2019) Interplay of protein corona and immune cells controls blood residency of liposomes. Nat Commun 10(1):1–11

Hajipour MJ, Ghasemi F, Aghaverdi H, Raoufi M, Linne U, Atyabi F, Bargahi A (2017) Sensing of Alzheimer's disease and multiple sclerosis using nano-bio interfaces. J Alzheimers Dis 59 (4):1187–1202

Jin G, He R, Liu Q, Dong Y, Lin M, Li W, Xu F (2018) Theranostics of triple-negative breast cancer based on conjugated polymer nanoparticles. ACS Appl Mater Interfaces 10(13):10634–10646

Joerger M (2012) Prevention and handling of acute allergic and infusion reactions in oncology. Ann Oncol 23(10):313–319

Johansson SG (2001) A revised nomenclature for allergy an EAACI position statement from the EAACI nomenclature task force. Allergy 56(9):813–824

Jutel M (2015) International consensus on allergy immunotherapy. J Allergy Clin Immunol 136 (3):556–568

Kah JCY, Yeo ELL, Koh WL, Poinard BEA, Neo DJH (2013) Nanoparticle Interface to biology: applications in probing and modulating biological processes. Crit Rev Biomed Eng 41 (4–5):323–341

Kah JC, Grabinski C, Untener E, Garrett C, Chen J, Zhu D, Hussain SM, Hamad-Schifferli K (2014) Protein coronas on gold Nanorods passivated with Amphiphilic ligands affect cytotoxicity and cellular response to penicillin/streptomycin. ACS Nano 8(5):4608–4620

Ke PC, Lin S, Parak WJ, Davis TP, Caruso F (2017) A decade of the protein corona. ACS Nano 11 (12):11773–11776

Khan DA (2016) Hypersensitivity and immunologic reactions to biologics: opportunities for the allergist. Ann Allergy Asthma Immunol 117(2):115–120

Kharazian B, Lohse SE, Ghasemi F, Raoufi M, Saei AA, Hashemi F, Hadipour NL (2018) Bare surface of gold nanoparticle induces inflammation through unfolding of plasma fibrinogen. Sci Rep 8(1):12557–12565

Kim Y, Ko SM, Nam JM (2016) Protein-nanoparticle interaction-induced changes in protein structure and aggregation. Chem Asian J 11(13):1869–1877

Konar M, Mathew A, Dasgupta S (2019) Effect of silica nanoparticles on the amyloid fibrillation of lysozyme. ACS Omega 4(1):1015–1026

Kozma GT, Mészáros T, Weiszhár Z, Schneider T, Rosta A, Urbanics R, Szebeni J (2015) Variable association of complement activation by rituximab and paclitaxel in cancer patients in vivo and in their screening serum in vitro with clinical manifestations of hypersensitivity: a pilot study. Eur J Nanomed 7(4):289–301

Kroschinsky F, Stölzel F, von Bonin S, Beutel G, Kochanek M, Kiehl M, Schellongowski P (2017) New drugs, new toxicities: severe side effects of modern targeted and immunotherapy of cancer and their management. Crit Care 21(1):89–99

Le QV, Yang G, Wu Y, Jang HW, Shokouhimehr M, Oh YK (2019) Nanomaterials for modulating innate immune cells in cancer immunotherapy. Asian J Pharm Sci 14(1):16–29

Ledford H (2016) Bankruptcy filing worries developers of nanoparticle cancer drugs. Nature 533 (7603):304–305

Lee DW, Gardner R, Porte DL, Louis CU, Ahmed N, Jensen M, Mackall CL (2014) Current concepts in the diagnosis and management of cytokine release syndrome. Blood 124 (2):188–195

Lee YK, Choi EJ, Webster TJ, Kim SH, Khang D (2015) Effect of the protein corona on nanoparticles for modulating cytotoxicity and immunotoxicity. Int J Nanomedicine 10:97–113

Lenz HJ (2007) Management and preparedness for infusion and hypersensitivity reactions. Oncologist 12(5):601–609

Li J, Huang J, Ao Y, Li S, Miao Y, Yu Z, Yang X (2018) Synergizing upconversion nanophotosensitizers with hyperbaric oxygen to remodel the extracellular matrix for enhanced photodynamic cancer therapy. ACS Appl Mater Interfaces 10(27):22985–22996

Liu W, Rose J, Plantevin S, Auffan M, Bottero JY, Vidaud C (2013) Protein corona formation for nanomaterials and proteins of a similar size: hard or soft corona? Nanoscale 5(4):1658–1668

Lorenzo-Abalde S, Gonzalez-Fernandez A (2013) In: Dobrovolskaia MA, McNeil SE (eds) Handbook of immunological properties of engineered nanomaterials. vol 1. World Scientific, Singapore, pp 517–546

Lundqvist M, Sethson I, Jonsson H (2004) Protein adsorption onto silica nanoparticles: conformational changes depend on the particles' curvature and the protein stability. Langmuir 20 (24):10639–10647

Lundqvist M, Stigler J, Cedervall T, Elia G, Lynch I, Dawson K (2008) Nanoparticle size and surface properties determine the protein Corona with possible implications for biological impacts. Proc Natl Acad Sci U S A 105(38):14265–14270

Lundqvist M, Stigler J, Cedervall T, Berggård T, Flanagan MB, Lynch I, Elia G, Dawson K (2011) The evolution of the protein Corona around nanoparticles: a test study. ACS Nano 5 (9):7503–7509

Lynch I, Dawson KA (2008) Protein-nanoparticle interaction. Nano Today 3(1–2):40–47

Lynch I, Salvati A, Dawson KA (2009) Protein-nanoparticle interactions: what does the cell see? Nat Nanotechnol 4(9):546–547

Ma L, Yang C, Huang L, Chen Y, Li Y, Cheng C, Cheng B, Zheng K (2019) Glycated insulin exacerbates the cytotoxicity of human islet amyloid polypeptides: a vicious cycle in type 2 diabetes. ACS Chem Biol 14(3):486–496

Mahmoudi M (2018) Debugging nano–bio interfaces: systematic strategies to accelerate clinical translation of nanotechnologies. Trends Biotechnol 36(8):755–769

Mahmoudi M, Kalhor H, Laurent S, Lynch I (2013) Protein fibrillation and nanoparticle interactions: opportunities and challenges. Nanoscale 5(7):2570–2588

Markman M, Kennedy A, Webster K, Kulp B, Peterson G, Belinson J (2000) Paclitaxel-associated hypersensitivity reactions: experience of the gynecologic oncology program of the Cleveland Clinic Cancer Center. J Clin Oncol 18(1):102–102

Meisel CL, Bainbridge P, Mitsouras D, Wong JY (2018) Targeted nanoparticle binding to hydroxyapatite in a high serum environment for early detection of heart disease. ACS Appl Nano Mater 1(9):4927–4939

Milani S, Baldelli Bombelli F, Pitek AS, Dawson KA, Rädler J (2012) Reversible versus irreversible binding of transferrin to polystyrene nanoparticles: soft and hard Corona. ACS Nano 6 (3):2532–2541

Mirshafiee V, Kim R, Park S, Mahmoudi M, Kraft ML (2016) Impact of protein pre-coating on the protein corona composition and nanoparticle cellular uptake. Biomaterials 75:295–304

Moghimi SM (2018) Nanomedicine safety in preclinical and clinical development: focus on idiosyncratic injection/infusion reactions. Drug Discov Today 23(5):1034–1042

Monopoli MP, Walczyk D, Campbell A, Elia G, Lynch I, Bombelli FB, Dawson KA (2011) Physical-chemical aspects of protein corona: relevance to in vitro and in vivo biological impacts of nanoparticles. J Am Chem Soc 133(8):2525–2534
Monopoli MP, Åberg C, Salvati A, Dawson KA (2012) Biomolecular coronas provide the biological identity of nanosized materials. Nat Nanotechnol 7(12):779–786
Morgan BP, Harris CL (2015) Complement, a target for therapy in inflammatory and degenerative diseases. Nat Rev Drug Discov 14(12):857–877
Nel AE, Mädler L, Velegol D, Xia T, Hoek EM, Somasundaran P, Thompson M (2009) Understanding biophysicochemical interactions at the nano–bio interface. Nat Mater 8(7):543–557
Nguyen VH, Lee B-J (2017) Protein corona: a new approach for nanomedicine design. Int J Nanomedicine 12:3137–3151
Nierenberg D, Khaled AR, Flores O (2018) Formation of a protein corona influences the biological identity of nanomaterials. Rep Pract Oncol Radiothe 23(4):300–308
Obst K, Yealland G, Balzus B, Miceli E, Dimde M, Weise C, Charbaji N (2017) Protein corona formation on colloidal polymeric nanoparticles and polymeric nanogels: impact on cellular uptake, toxicity, immunogenicity, and drug release properties. Biomacromolecules 18 (6):1762–1771
Oh JY, Kim HS, Palanikumar L, Go M, Jana B, Park SA, Kim C (2018) Cloaking nanoparticles with protein corona shield for targeted drug delivery. Nat Commun 9(1):4548–4556
Panet E, Mashriki T, Lahmi R, Jacob A, Ozer E, Vecsler M, Tzur A (2017) The interface of nanoparticles with proliferating mammalian cells. Nat Nanotechnol 12(7):598–600
Papi M, Palmieri V, Digiacomo L, Giulimondi F, Palchetti S, Ciasca G, Coppola R (2019a) Converting the personalized biomolecular corona of graphene oxide nanoflakes into a high-throughput diagnostic test for early cancer detection. Nanoscale 11(32):15339–15346
Papi M, Palmieri V, Palchetti S, Pozzi D, Digiacomo L, Guadagno E, Mahmoudi M (2019b) Exploitation of nanoparticle-protein interactions for early disease detection. Appl Phys Lett 114 (16):163702–163707
Patel SV, Khan DA (2017) Adverse reactions to biologic therapy. Immunol Allergy Clin N Am 37 (2):397–412
Pelaz B, del Pino P, Maffre P, Hartmann R, Gallego M, Rivera-Fernandez S, Parak WJ (2015) Surface functionalization of nanoparticles with polyethylene glycol: effects on protein adsorption and cellular uptake. ACS Nano 9(7):6996–7008
Rahman M, Laurent S, Tawil N, Yahia LH, Mahmoudi M (2013) Protein corona: applications and challenges in protein-nanoparticle interactions. Springer, Berlin/Heidelberg, pp 45–63
Rao L, Meng QF, Bu LL, Cai B, Huan Q, Sun ZJ, Wang TH (2017) Erythrocyte membrane-coated upconversion nanoparticles with minimal protein adsorption for enhanced tumor imaging. ACS Appl Mater Interfaces 9(3):2159–2168
Ritz S, Schöttler S, Kotman N, Baier G, Musyanovych A, Kuharev J, Mailänder V (2015) Protein corona of nanoparticles: distinct proteins regulate the cellular uptake. Biomacromolecules 16 (4):1311–1321
Rodriguez PL (2013) Minimal 'self' peptides that inhibit phagocytic clearance and enhance delivery of nanoparticles. Science 339(6122):971–975
Sampson HA (2006) Second symposium on the definition and management of anaphylaxis: summary Report Second National Institute of Allergy and Infectious Disease/Food Allergy and Anaphylaxis Network symposium. J Allergy Clin Immunol 117(2):391–397
Sampson HA, Munoz-Furlong A, Campbell RL, Adkinson NF Jr, Bock SA, Branum A (2006) Second symposium on the definition and management of anaphylaxis: summary report– Second National Institute of Allergy and Infectious Disease/Food Allergy and Anaphylaxis Network symposium. J Allergy Clin Immunol 117(2):391–397
Saptarshi SR, Duschl A, Lopata AL (2013) Interaction of nanoparticles with proteins: relation to bio-reactivity of the nanoparticle. J Nanobiotechnol 11(1):1–12
Scarlet C (2006) Anaphylaxis. J Infus Nurs 29(1):39–44

Schöttler S, Becker G, Winzen S, Steinbach T, Mohr K, Landfester K, Wurm FR (2016) Protein adsorption is required for stealth effect of poly (ethylene glycol)-and poly (phosphoester)-coated nanocarriers. Nat Nanotechnol 11(4):372–377
Simons FE, Ebisawa M, Sanchez-Borges M, Thong BY, Worm M, Tanno LK, Sheikh A (2015) Update of the evidence base: World Allergy Organization anaphylaxis guidelines. World Allergy Organ J 8(1):32
Singla R, Abidi SMS, Dar AI, Acharya A (2019) Inhibition of Glycation-induced aggregation of human serum albumin by organic–inorganic hybrid nanocomposites of Iron oxide-functionalized Nanocellulose. ACS Omega 4(12):14805–14819
Su J, Sun H, Meng Q, Zhang P, Yin Q, Li Y (2017) Enhanced blood suspensibility and laser-activated tumor-specific drug release of theranostic mesoporous silica nanoparticles by functionalizing with erythrocyte membranes. Theranostics 7(3):523–537
Szebeni J (2005) Complement activation-related pseudoallergy: a new class of drug-induced acute immune toxicity. Toxicology 216(2–3):106–121
Szebeni J (2012a) A porcine model of complement-mediated infusion reactions to drug carrier nanosystems and other medicines. Adv Drug Deliv Rev 64(15):1706–1716
Szebeni J (2012b) Hemocompatibility testing for nanomedicines and biologicals: predictive assays for complement mediated infusion reactions. Eur J Nanomed 4(1):33–53
Szebeni J, Muggia F, Gabizon A, Barenholz Y (2011) Activation of complement by therapeutic liposomes and other lipid excipient-based therapeutic products: prediction and prevention. Adv Drug Deliv Rev 63(12):1020–1030
Szebeni J, Bedőcs P, Dézsi L, Urbanics RA (2018a) Porcine model of complement activation-related pseudoallergy to nanopharmaceuticals: pros and cons of translation to a preclinical safety test. Pre Nanomed 2:63–72
Szebeni J, Simberg D, González-Fernández Á, Barenholz Y, Dobrovolskaia MA (2018b) Roadmap and strategy for overcoming infusion reactions to nanomedicines. Nat Nanotechnol 13 (12):1100–1108
Talapin DV, Shevchenko EV (2016) Introduction: nanoparticle chemistry. Chem Rev 116:10343–10345
Tanokashira D, Mamada N, Yamamoto F, Taniguchi K, Tamaoka A, Lakshmana MK, Araki W (2017) The neurotoxicity of amyloid β-protein oligomers is reversible in a primary neuron model. Mol Brain 10(1):1–10
Teichroeb JH, Forrest JA, Jones LW (2008) Size-dependent denaturing kinetics of bovine serum albumin adsorbed onto gold nanospheres. Eur Phys J E Soft Matter 26(4):411–415
Tenzer S, Docter D, Kuharev J, Musyanovych A, Fetz V, Hecht R, Schlenk F, Fischer D, Kiouptsi K, Reinhardt C (2013) Rapid formation of plasma protein Corona critically affects nanoparticle pathophysiology. Nat Nanotechnol 8(10):272–281
Treuel L, Brandholt S, Maffre P, Wiegele S, Shang L, Nienhaus GU (2014) Impact of protein modification on the protein corona on nanoparticles and nanoparticle–cell interactions. ACS Nano 8(1):503–513
Walczyk D, Baldelli Bombelli F, Monopoli MP, Lynch I, Dawson K (2010) What the cell "Sees" in bionanoscience. J Am Chem Soc 132(16):5761–5768
Walkey CD, Chan WC (2012) Understanding and controlling the interaction of nanomaterials with proteins in a physiological environment. Chem Soc Rev 41(7):2780–2799
Wells MA, Abid A, Kennedy IM, Barakat AI (2012) Serum proteins prevent aggregation of Fe2O3 and ZnO nanoparticles. Nanotoxicology 6(8):837–846
Wetterskog E, Tai CW, Grins J, Bergström L, Salazar-Alvarez G (2013) Anomalous magnetic properties of nanoparticles arising from defect structures: topotaxial oxidation of Fe1–x O|Fe3–δO4 Core|shell nanocubes to single-phase particles. ACS Nano 7(8):7132–7144
Wolfram J, Suri K, Celia C, Yang Y, Shen J, Fresta M, Zhao Y, Shen H (2013) Ferrari M shrinkage of pegylated and non-pegylated liposomes in serum. Colloids Surf B Biointerfaces 1 (114):294–300

Wolfram J, Yang Y, Shen J, Moten A, Chen C, Shen H, Zhao Y (2014) The nano-plasma interface: implications of the protein corona. Colloids Surf B: Biointerfaces 124:17–24

Yan Y, Gause KT, Kamphuis MM (2013) Differential roles of the protein corona in the cellular uptake of nanoporous polymer particles by monocyte and macrophage cell lines. ACS Nano 7 (12):10960–10970

Zeng Z, Zhao P, Liu L, Gao X, Mao HQ, Chen Y (2018) Lipid stabilized solid drug nanoparticles for targeted chemotherapy. ACS Appl Mater Interfaces 10(30):24969–24974

Zhang Z, Guan J, Jiang Z, Yang Y, Liu J, Hua W, Zhan C (2019) Brain-targeted drug delivery by manipulating protein corona functions. Nat Commun 10(1):1–11

Zheng T, Finn C, Parrett CJ, Dhume K, Hwang JH, Sidhom D, Huo Q (2017) A rapid blood test to determine the active status and duration of acute viral infection. ACS Infect Dis 3(11):866–873

Critical Overview of the Subject: Current Scenario and Future Prospects 9

Pooja, Sarthak Sharma, Avnesh Kumari, and Amitabha Acharya

Contents

Pooja · S. Sharma
Biotechnology Division, CSIR-Institute of Himalayan Bioresource Technology, Palampur, Himachal Pradesh, India

A. Kumari · A. Acharya (✉)
Biotechnology Division, CSIR-Institute of Himalayan Bioresource Technology, Palampur, Himachal Pradesh, India

Academy of Scientific & Innovative Research (AcSIR), CSIR-Institute of Himalayan Bioresource Technology, Palampur, Himachal Pradesh, India
e-mail: amitabha@ihbt.res.in

A. Acharya (ed.), *Nanomaterial - Based Biomedical Applications in Molecular Imaging, Diagnostics and Therapy*, https://doi.org/10.1007/978-981-15-4280-0_9

Abstract

Nanomaterials (NMs) play a vital role in molecular imaging, disease diagnosis, and therapy because of their unique chemical and physical properties. The importance of nanomaterials in biomedical applications cannot be overlooked as they are the potential carrier and delivery agents of imaging probes, therapeutic drugs, or active biological material to desired target sites. Nanomaterial-based biomedicines have significant advantages over traditional or conventional ones such as better sensitivity, high target specificity, bioavailability, and biodistribution. In this chapter we have given a brief account on the advancements of different types of nanomaterials for their use in the molecular and bioimaging of a number of dreadful diseases including cancers, cardiovascular diseases, and neurodegenerative diseases. This chapter also gives details on the use of these nanomaterials in disease detection and therapy at the preclinical as well as clinical aspects. Finally, we have concluded with the future outlook for the proper use of these nanomaterials in molecular imaging, diagnostics, and therapy.

Keywords

Nanomaterials · Bioimaging · Disease detection · Therapy · Biomedical application

Abbreviations

CNTs	Carbon nanotubes
CT	Computed tomography
GCE	Glassy carbon electrode
IVDs	In vitro diagnostics
MAB	Monoclonal antibody
MRI	Magnetic resonance imaging
MRS	Magnetic resonance spectroscopy
MWCNTs	Multiwalled carbon nanotubes
NIR	Near infrared
NMs	Nanomaterials
NPs	Nanoparticles
PA	Photoacoustic
PDT	Photodynamic therapy
PEG	Polyethylene glycol
PET	Positron emission tomography
PPTT	Plasmonic photothermal therapy
PTT	Photothermal therapy

QDs	Quantum dots
SPECT	Single-photon emission computed tomography
SPIONs	Super-paramagnetic iron oxide nanoparticles
SPR	Surface plasmon resonance
SWCNTs	Single-walled carbon nanotubes
UCNPS	Upconversion nanoparticles

9.1 Introduction

Molecular imaging has become one of the most fascinating and rapidly emerging areas in the field of science. It provides insight for understanding the underlying biology, early diagnosis of various diseases, and assessment of the therapeutic responses. From decades, there have been significant advancements in a number of imaging tools and their applications, aiming to establish a new generation of molecular imaging platforms with better sensitivity and specificity for in vivo visualization and evaluation of various biological processes. Molecular imaging creates a platform for amplification and quantification of signals generated by the probe at the target site, but getting a high contrast signal under in vivo conditions is the most important limiting factor, because of the interaction of a large number of biomacromolecules present in the biological milieu. Further, there are a number of hurdles that are associated with the NPs, viz., stability, toxicity, and also cellular compatibility. Moreover, for improving the specificity and targetability of the molecular imaging probes, they are usually conjugated with specific moieties such as fluorescent probes, antibodies, peptides, aptamers, etc.

Early diagnostics is a key for efficient and effective treatment. Generally, early testing is far economical and suitable than the treatments done after actual occurrence of the disease. Keeping this in mind, there has been an increasing demand for cheaper, sensitive, selective, accurate, stable, and faster diagnostic systems. Current research advancements on nanotechnology have exploited the ways to synthesize different types of NMs, which can be significantly used for diagnostic purposes. NMs due to their inherent properties have a wide potential to serve as an ideal material for diagnostics. NMs possess larger active surface area which makes them highly specific and sensitive for any application; moreover, there is the ease for surface modifications (Mitragotri et al. 2015). These properties are being exploited to make NMs suitable for diagnostics. There are two main approaches that are generally used to synthesize NMs for diagnostics purposes. First is the top-down approach, where NMs are fabricated from the bulk, and second is the bottom-up approach, where NMs are prepared by assembling the basic units to form a larger unit of desired dimension. Regardless of the approach used, the aim is to synthesize the NMs with controlled size and significant stability, which could be suited for better diagnosis. The surface coating of the NMs plays an important role as per size and stability concerns. Multidentate small ligand moieties can be the facile approach for achieving the stable surface with controlled size (Ma et al. 2016; Zhan et al. 2013). Another issue with respect to highly charged NMs is its nonspecific binding

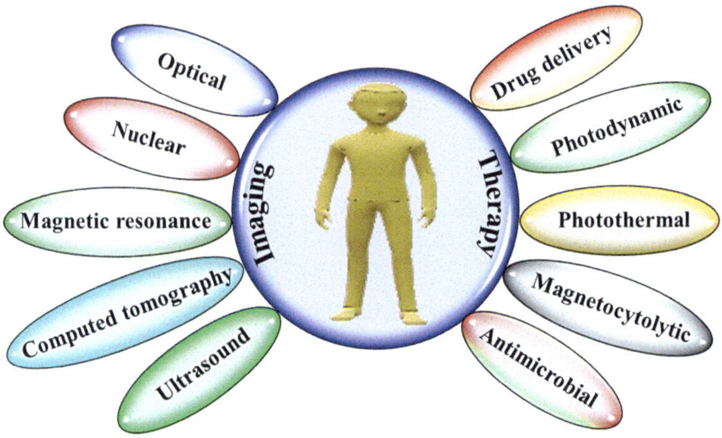

Fig. 9.1 Figure showing different types of imaging modalities and therapeutic approaches in the NM-based future medicine

to the biomarker which may lead to the change in various physiochemical properties of the NMs such as hydrodynamic size, which results in poor sensitivity in diagnostics. Surface coating of NMs with polyethylene glycol (PEG) has been a facile approach extensively used for reducing the charge and nonspecific binding while maintaining the size and colloidal stability of NMs. There are various types of NMs which are being used in in vitro and in vivo diagnostics. In in vitro diagnosis, samples are taken out of the body and then tested further, whereas in in vivo diagnosis, the test is carried out within the body. Current advancements in nanotechnology and in vitro diagnostics (IVDs) have been encouraging the patients for self-diagnosis with more sensitivity and specificity. There are several approaches that could be employed for diagnostics with NMs, viz., fluorescence, Raman scattering, magnetic resonance imaging (MRI), positron emission tomography (PET), and computed tomography (CT) (Fig. 9.1).

The use of nanoparticles in medicines is widely applicative for therapy of several diseases because of various advantages that overcome the major limitations faced by traditional or conventional therapies, viz., bioavailability, low target specificity, and different side effects. Introductory reports of nanoparticles that were used in disease therapy involved an inert nanomaterial that used as a vector of various classes drug molecules, viz., anesthetics, chemotherapeutics, immunosuppressants, steroids, antibiotics, vaccines, etc., for efficient delivery and therapy. With the growing interest of researchers in nanomaterial-based biomedicines, novel types of NMs were synthesized which can be used as drugs themselves because of their unique physical and surface properties. With the advancements in nanomaterial-based bioimaging and diagnostics of various diseases, NMs are emerging as a promising tool for the therapy of various diseases.

9.2 Current Scenario of Nanoparticles in Molecular Imaging

Presently, a plethora of nanoparticles are being explored for biomedical applications with the objective of molecular imaging, early diagnosis, and better therapy of severe diseases. Engineered nanoparticles are exponentially investigated in the field of molecular imaging due to unique chemical (high reactivity and stability) and physical (high surface-to-volume ratio) properties which emerge only at nanoscale levels. A variety of nanoparticles such as quantum dots, metallic nanoparticles, nanotubes, solid lipid nanoparticles, polymeric micelles, liposomes, dendrimers, polymeric nanoparticles, and iodinated nanoparticles are under examination for biomedical molecular imaging. Currently, there are numerous imaging procedures for molecular imaging, viz., computed tomography (CT), magnetic resonance imaging (MRI), magnetic resonance spectroscopy (MRS), ultrasound (US), positron emission tomography (PET), and single-photon emission computed tomography (SPECT). PET, SPECT, and MRS are major imaging modalities that are routinely operated in clinics for medical imaging purposes, including cardiology, oncology and neurology. MRS measures proton excitation and relaxation which is directly linked with metabolites, viz., polyamines, lipids, lactates, pyruvates, cholines, and much more. Low penetration depth in human tissue is one of the major limiting factors associated with optical imaging under in vivo conditions. There are a number of technological advancements in a catheter (e.g., imaging of bladder cancer) and endoscopic devices (e.g., investigating Barrett's esophagus) for ocular coherence tomography, which are promising tools for novel clinical applications. In this section, we will focus on the present NP-mediated molecular imaging techniques and modalities along with their status in clinics.

9.2.1 Nanoparticles in Magnetic Resonance Imaging

Nanoparticles play an important role in MR-based imaging probes, being able to produce contrasts (both positive and negative) in T1/T2-weighted images. They also help in increasing the contrast by the alternation of water proton relaxivity. Iron oxide NPs having superparamagnetic nature (SPIONs) are the first used nanoparticles in biomedical imaging in animal studies. SPIONS are used as good contrast agents which develop the localized dark region in MRI images in the case of T2-weighted images. Most of the SPIONs have been used for clinical trials as in the case of cancer patients, including ferumoxytol (Bashir et al. 2015), ferumoxtran (Bernd et al. 2009; Heesakkers et al. 2009), ferucarbotran NPs, etc. (Rief et al. 2009). Ferumoxytol is used as an "off-label" contrast agent in 20 current ongoing clinical trials as a detector of different diseases simultaneously. Ferumoxtran-10 is currently undergoing clinical trials in Europe (NCT02751606, NCT03223064, NCT02549898, NCT02997046) (Daldrup-Link 2017). Moreover, different types of nanoparticles have been explored in preclinical studies for contrast-enhanced imaging such as SPION-mediated noninvasive detection of elevated mRNA level after long-term drug exposure (Liu et al. 2009), tracking the expression of genes in rodent brains after gene therapy (Ren et al. 2016), theranostic applications of magnetic iron oxide NPs in cancer (Thomas et al. 2013), etc.

9.2.2 Nanoparticles in Optical Imaging

The remarkable optical imaging, having a high spatial resolution, possesses limited tissue penetration because of the signal attenuation in the tissues in this wavelength range. This limiting factor allows them to image tissues superficially. Most of the NPs possess inherent fluorescence and are suitable for ocular or near-infrared (NIR) imaging (Wolfbeis 2015). Preclinical studies make use of photoluminescent NPs as they have polished surface functionalization and ocular properties. In order to improve tissue penetration with less tissue damage, negligible autofluorescence, and multicolor capabilities, upconverting nanoparticles (UCNPs) doped with lanthanide were designed for imaging under in vivo conditions and photothermal therapy (PTT) (Chen et al. 2014; Lv et al. 2017). Since the last 20 years, QDs due to their versatile spectra and better quantum yield have been explored for fluorescent imaging of tissues. Cadmium is present in typical quantum dots, which is hazardous to human health and prevents their clinical usage (Rzigalinski and Strobl 2009; Xu et al. 2016). Therefore, the quest for the synthesis of novel fluorescent material is underway.

9.2.3 Nanomaterial-Based Ultrasound Imaging

Although ultrasound imaging is a potential imaging modality for the detection of severe malignancies due to its low cost, absence of ionizing radiations, and accessibility, its use is limited in cancer applications because US imaging is limited to distinguishable scattering properties and tissue boundaries differentiation. Ultrasound imaging makes use of microbubbles and their alterations as the contrast agents. As microbubbles having large size range (1–8 μm), they are confined to vasculature imaging, and for this reason, they had limited use in oncology (Gao et al. 2017). In the same study, it was demonstrated that better diagnosis of epithelial ovarian cancer can be performed through US molecular imaging by cancer antigen 125 (CA-125)-targeted nanobubbles. It has been clearly seen that with the advancement and evolvement in nanomaterials, US molecular imaging can prevail abovementioned limitations of microbubbles.

9.2.4 Nanomaterial-Based Computed Tomography (CT)

CT has been extensively used in imaging modality for the screening of cancer due to its ample availability and comparatively low cost. CT can readily visualize the contrast of bone and lung tissue, yet soft tissue's molecular imaging utilizes high Z-contrast agents. Iodine-containing small molecules usually behave as contrast agents of CT, but these contrast agents undergo various toxicity concerns and exhibit short half-lives. These concerns can be overcome by the use of NPs, by introducing gold or iodine into a nanoparticle structure that enhances the stability, targeting specificity toward tumor site, and circulation for a longer period of time (Ashton et al. 2015). There are several nanomaterials such as dendrimers (Curry et al. 2014), iodine-containing liposomes (Badea et al. 2012), gold nanorods and shells, and lanthanide oxide NPs (Park et al. 2015) which have been extensively discovered for CT-based oncological imaging. Furthermore, nanomaterial-based contrast agents include micelle and used to mark extensively expressed molecule during disease amelioration (e.g., atherosclerosis, inflammation, and cancer) (Cormode et al. 2014). Despite having such advancements in the field of nanomaterial-based CT contrast agents, still, CT is limited to the preclinical level.

9.2.5 NPs in Nuclear Imaging Techniques

Nuclear imaging techniques that mainly include PET and SPECT due to their instinctive sensitivity offer extreme advancements in clinics. The techniques such as PET and SPECT provide unprecedented and unique sensitivity for the detection for the cells/tissues at a much deeper level as compared to the other type of imaging modalities. The radiolabeling of these NPs can be done using chelators or by intrinsic labeling with radionuclides, which can be further employed for evaluation of biodistribution and studying tumor targeting. With this excellent and unique tunability in designing and manipulation, this field of nuclear medicine has made great progress in the present era of the biomedical research field.

9.2.6 Photoacoustic Imaging

In photoacoustic imaging, there is the utilization of sound waves to generate an image as in US imaging, but apart from that PA imaging makes use of NIR, i.e., "light-in," for the purpose of tissue irradiation and detecting the resulting pressure waves, i.e., "sound off" (Zackrisson et al. 2014). PA provides anatomical structure and flow rates along with clue about ocular, mechanical, and physiological properties of the tissue. Organic as well as inorganic NPs are being used as contrast agents in PA imaging preclinically. NPs having inorganic framework as in the case of metallic NPs, QDs (Shashkov et al. 2008), and UCNPs (Maji et al. 2014) have been successfully explored preclinically for the visualization of cancer lesions. Similarly, NPs having organic framework such as carbon NPs have been used in PA imaging studies (Weber et al. 2016).

9.3 Current Scenario of Nanomaterials in Disease Diagnosis

9.3.1 Metal-Based NMs in Disease Diagnosis

Metal-based NMs have gained a significant interest among researchers for point-of-care infectious disease diagnostics due to their sensitivity and specificity, leading to efficient results with reduced transmissions. For improving the sensitivity, metal-based sample preparation and metal-based signal amplification could be done (Markwalter et al. 2018). A plethora of studies have been done on the noble-metal NM-based in vitro diagnostic systems. Due to their unique and unprecedented applicative properties, viz., amplified localized surface plasmon resonance (LSPR), fluorescence resonance energy transfer (FRET), and also the size-dependent Raman scattering properties, these NMs can be used for point-of-care (POC) in vitro diagnostics in healthcare sector (Howes et al. 2014; Zhou et al. 2015). Many studies relating to biosensing diagnostic applications of the engineered metal-based NMs with controlled size have been done. Physical parameters such as size and shape of NMs could significantly impact the signal transduction (Xi et al. 2018). A study has reported the use of sequential pyramids of silver nanoparticles (AgNPs) as Raman reporter molecules when attached with multiple aptamers (Xu et al. 2015). In another study, AuNPs (gold nanoparticles) coupled with AgNPs in the form of nanodumbbells were reported to have specific single-molecule detection ability based on SERS sensing (Lim et al. 2010). Similar sensing approach has been carried out by using AuNP-on-wire sensing systems for diagnosis of various pathogens and other biosensing applications (Kang et al. 2010). Chiroplasmonic properties of a heterodimer of gold and silver NPs with prolate geometries have been harnessed for the detection of various biomarkers (Wu et al. 2013). Rica et al. reported colorimetric detection of disease biomarker via AuNP-mediated conventional enzyme-linked immunosorbent assay (ELISA). Enzymes linked in ELISA control the growth of AuNPs and produce red and blue colors in the absence and presence of analyte, respectively (De La Rica and Stevens 2012). In a similar study, colorimetric detection of horseradish peroxidase (HRP) enzyme molecules linked to the single plasmonic AuNPs has been made for the detection of low-abundance biomarkers (Chen et al. 2011). Ultrasensitive diagnosis of H5N1 and ALP virus was also carried out by vivid colorimetric signals produced by the deposition of AuNPs on gold nanobipyramids (AuNBPs) (Xu et al. 2017a, b). Concentration of the analytes plays a major role in the signal transmission. Ultra-low concentration of the analytes makes it difficult for the conventional transducers to generate desired signals; hence, such diagnosis lacks credibility. For increasing the signal strength at low concentrations, a glucose oxidase was conjugated to AgNPs on Au nanosensors (Rodríguez-Lorenzo et al. 2012). In another study, low-concentration DNA detection has been carried out by the etching of triangular silver nanoprism to a different shape, which resulted in a significant shift in surface plasmonic resonance signals (Yang et al. 2014). Heterogeneously coupled NMs have also gathered much attention in the field of point-of-care diagnostics. Catalytic and magnetic properties of the Pt-functionalized Fe_3O_4 bioassay have been harnessed for creating synergetic IVD systems with considerable sensitivity (Kim et al. 2017a, b). In another study, enhancement in catalytic activity was achieved by in situ synthesized PdNPs decorated ultrathin $NiCl_2$ nanosheets (NSs). NSs substantially catalyze the rate of oxidation of tetramethylbenzidine (Cai et al. 2018). Approaches such as ELISA, cell cultures, and polymerase chain reaction (PCR) technology have been facing issues due to their complex, onerous, and expensive nature (Inci et al. 2013). Inci et al. have reported a facile strategy for detection and quantification of multiple HIV subtypes by label-free nanoplasmonic detection system. Despite the encouraging efforts done in the field of nanotechnology and metal-based NMs, there are some real issues regarding the reproducibility of these NMs at the clinic level.

9.3.2 Carbon-Based NMs in Diagnostics

Carbon nanomaterials in various forms, viz., fullerenes, carbon nanotubes (CNTs), graphenes, carbon-based quantum dots, and nanodiamonds, serve as potential materials for medical diagnostics. Due to the fascinating physiochemical, electrochemical, and optical properties of these NMs, they find application in various biosensing and imaging systems, starting with fullerenes (C_{60}), a supramolecule known for its highly symmetrical geometry that can be hollow sphere, ellipsoid, or tube comprising of carbon atoms. C_{60} NPs are insoluble in water, but due to its ease of surface fictionalization, hydrophilic moieties can be readily attached, increasing its solubility and biocompatibility (Hu et al. 2008; 2010a, b). The tunable surface chemistry of these NMs makes it a potential candidate as a contrast agent in magnetic resonance imaging (MRI) for cancer diagnostics (Mody et al. 2009; Shu et al. 2009).

Carbon nanotubes (CNTs), another carbon-based NM which has number of applications in medical diagnostics. CNTs exist in two forms, viz., single-walled carbon nanotubes (SWCNTs) and multiwalled carbon nanotubes (MWCNTs). CNTs have fascinating optical and physical properties, which can be harnessed in the form of contrast agent and various CNT-based biosensors. Several studies have reported the use of CNTs for detection of prostate-specific antigen (PSA), which is a biomarker for prostate cancer (Feng et al. 2017; Tian et al. 2012). In another study, electrochemical biosensing of glucose has been achieved by glucose oxidase-coated MWCNTs. Emphasis was also given on how the structure and surface chemistry affect the performance of the biosensor. The study concluded that the immobilization of glucose oxidase had been significant in the case of oxidized CNTs (González-Gaitán et al. 2017). Similar electrochemical biosensing strategy has been employed on near-infrared fluorescent SWCNTs for the spatial imaging of nitric oxide that could be used in diagnosis of endothelial dysfunction for the study of tumor progression (Ulissi et al. 2014). Monoclonal antibody (MAB)-coated MWCNTs were employed for the biosensing of cardiovascular disorders (CVDs). The synthesized immunosensor is based on the impedance measurements, which selectively and sensitively detected the presence of myoglobin (Khan et al. 2016). Navamani et al. prepared a nanoprobe for the detection of blood cholesterol by coating a film containing Nafion, MWCNT, cholesterol oxidase, and poly-N-vinyl-2-pyrrolidone encapsulated ZnS NPs onto GCE (glassy carbon electrode) (Navamani et al. 2012). In another study, potentiometric detection of dopamine present in the blood serum was carried out using MWCNT-based molecularly imprinted polymer (Anirudhan et al. 2014). CNTs also offer a noninvasive sensing strategy for volatile organic compound (VOC) sensing (Ellis and Star 2016; Tang et al. 2015).

Graphenes are two-dimensional hexagonal array of carbon atoms. Due to its unique tunable properties, it has been emerged as prime NMs in the field of biosensing (Patil et al. 2015; Ghosal and Sarkar 2018). Graphene NMs have been extensively used in the glucose sensing (Fu et al. 2009), where glucose oxidase-coated graphene-Nafion nanocomposites were used. In a similar study, electrochemical sensing of glucose has been done by chitosan-coated AuNP-graphene nanocomposite sheets (Shan et al. 2010). Diagnosis of diseases such as Alzheimer's and prostate cancer could also be done by mRNA sensing using graphene-upconversion NPs (Vilela et al. 2016). In another study, sensing of lactate biomarker has been done by using $PhNO_2$-functionalized reduced graphene oxide (Manna and Raj 2016). Graphene oxide serves as a potential Raman probe for far imaging. AuNP-functionalized graphene oxide has been used as an imaging tool for Hella229 cells (Liu et al. 2012). Photoacoustic imaging is another approach for noninvasive imaging. Lalwani et al. have substantially shown the potential of graphene oxide-based ribbons as contrasting agents in photoacoustic imaging (Lalwani et al. 2013).

9.3.3 Polymeric NMs in Detection of Various Diseases

Among the different types of NMs used for the diagnostic applications, polymeric NMs have shown a huge potential. Because of the unique properties such as excellent biocompatibility, biodegradability, and structural versatility, the theranostic applications of the polymeric NMs have been in the market in recent times (Clawson et al. 2011; Hu et al. 2010a, b). Among these, the naturally occurring biopolymers show safe degradability inside the body over a certain time period. These polymers are generally nontoxic except at very high concentrations. Polymers such as poly(D,L-lactic acid), polyethylene glycol (PEG), poly(D,L-glycolic acid), and polycaprolactone (PCL) are already there in the clinical trials. These NMs have already been taken into consideration for the treatment and diagnostics of a number of diseases. These polymeric NMs possess some other unique biophysicochemical properties, enabling them to increase their efficacy as compared to the other NMs. These also have shown applications in encapsulation of drugs as well as certain contrasting agents (Hu et al. 2014; Petros and DeSimone 2010; Wang et al. 2012). These NMs have prolonged circulation time inside the body, which helps them in sustained release, and also show accumulation at the specific sites through passive targeting by enhanced permeability and retention (EPR) effect. Many of these polymeric nanoformulations have now been incorporated to further enhance specific delivery to the tumor site for the sensitive and selective diagnostic, therapy, and imaging. Polymeric NMs have been utilized for different disease diagnostic biomedical applications. It provides advanced sensitive application (Ashley et al. 2018) and selective sensing platform (Oh et al. 2012). Recently, these NMs are under clinical trial for apoptosis imaging for image-guided therapy for selectively detecting real-time cancerous tissue from normal tissue (Kim et al. 2006). Flexible sensor based on polymeric nanomaterial is used for healthcare and environmental monitoring of different parameters (Segev-Bar and Haick 2013). Polymeric nanomaterials are also used for temperature sensing in living cells (Wang et al. 2020). Research on cancer-related diagnostics has been the prime importance for the past many years. Diagnosis of pancreatic adenocarcinoma has been laid out by using poly(methacrylic acid) (PMAAc)-based single-chain polymer NPs decorated with PTR86 peptide moieties by the help of single-photon emission computed tomography (SPECT) imaging (Benito et al. 2016). In another study, detection of tumor has been carried out with the help of semiconducting polymer NPs encrust with cell membranes of activated fibroblasts. These NPs are capable of giving substantially intense fluorescence and photoacoustic signals, an effective and efficient detection of tumor site (Li et al. 2018).

9.4 Current Scenario of Nanotherapeutics

9.4.1 Drug Delivery

NM-based drug delivery involves drug incorporation within nano-sized vector mainly via absorption, adsorption, conjugation, and encapsulation for stable dispensation in the body. These nanoparticles are generally known as nano-drug delivery systems (nano-DDS). These nano-DDS are well-known for efficient and targeted drug delivery. In addition, they can overcome the limitation of multidrug resistance effect which is mainly offered by tumor and diseased cells. Nano-DDS can play a vital role in combinational therapy as they can carry multiple drugs at the same time, as a single drug is less effective in the case of more complex diseases such as cancer. Initially, nanocarrier-based drug delivery moves around the incorporation of hydrophobic and small drug molecules, but now these techniques have expanded its horizon to another level which includes the incorporation of large molecules such as peptides, proteins, DNA, and hydrophilic molecules. Earlier polymers and lipid-based nano-DDS were used, but recently carbon-based porous inorganic nano-DDS have also been used. Many anticancer lipophilic drugs such as doxorubicin, camptothecin, and taxanes were encapsulated in self-assembled liposomal formulations. Doxil, a PEGylated liposomal formulation for doxorubicin drug against breast cancer (Kamaly et al. 2012). Similarly, AmBisome, a liposomal formulation for amphotericin-B, has been used as an antifungal agent (Moen et al. 2009). Again, CPX-351, a liposomal formulation, encapsulates cytarabine and daunorubicin drugs together and targets combinational therapy against amyloid leukemia (Feldman et al. 2011).

9.4.2 Photothermal Therapy (PTT)

PTT is also known as plasmonic photothermal therapy (PPTT) because PTT has been effectively used for localized plasmonic adsorption of metallic nanoparticles. In PTT, NP itself is used as a drug, and there is no need for organic photosensitizers (Kumar et al. 2014).

Customization of metallic NPs has been done in order to make them efficient to absorb electromagnetic NIR and IR light, which can generate local heating of approximately 113 °F. By localizing this type of NPs in tumor areas, immediate cell death and necrosis can be achieved as a result of light irradiation in that particular area. Gold nanoshells and gold nanorods have been efficiently used against thick tumors (Hirsch et al. 2003). Furthermore, gold nanorods have attracted more attention during recent years in the field of PTT because of their easy synthesis, small size, better stability, good absorption efficiency, and high surface plasmon resonance (Huang et al. 2009). Graphene oxide, reduced graphene oxide, and MoS_2 nanosheets are some assuring novel materials that are recently used as NIR PTT agents (Robinson et al. 2011; Chou et al. 2013).

9.4.3 Photodynamic Therapy (PDT)

PDT is a promising application to treat a variety of diseases including cancer, microbial infections and dermatological disorders (Verma et al. 2007; Morgan and Oseroff 2001). This is a light-activated therapeutic application in which photosensitizers are initially targeted to the disease site. Photosensitizers have a great affinity for malignant cells (due to EPR effect), therefore, their retention time increased, and thus get accumulated in the tumor. Their irradiation with light leads to transfer of energy to oxygen molecule or other molecules in the surrounding areas, generating cytotoxic reactive oxygen species, which can induce apoptotic and necrotic cell death (Dolmans et al. 2003). Low bioavailability, water solubility, biodistribution, and target specificity of photosensitizer drugs possess a major limitation with PDT. Therefore, organic moieties such as liposome-, polymeric- and carbon-based nanomaterials have been used to carry photosensitizer drugs (Hong et al. 2016). Inorganic nanomaterials such as gold NPs, silica NPs, and QDs have been reported to bind photosensitizer drugs (Colombeau et al. 2016; Cheng and Lo 2011).

9.4.4 Magnetocytolytic Therapy

An external magnetic field applied to magnetic NPs can lead to two of the most exclusive properties by which medicines can be benefited immensely (McCarthy and Weissleder 2008). Firstly, magnetic NPs are targeted to the diseased site under in vitro and in vivo condition by an externally applied magnetic field, permitting targeted drug delivery. Subsequently, activation of accumulated magnetic NPs was done by externally applied magnetic field to induce local heat (43–45 °C) for disease-specific toxicity which is known as magnetocytolysis, or magnetically induced hyperthermia (Dutz and Hergt 2014).

9.4.5 Antimicrobial Therapies

Metallic NPs are well-known for their toxicity toward pathogenic microbes, such as bacteria fungi, viruses, etc., because of their high surface energy. Silver is the most widely used metal as an antimicrobial agent as Ag^+ ions effectively inhibit the growth of microbes via suppression of various genes, enzymes, and electron transport units. Modern antimicrobial therapy using silver-containing creams is an important approach for the treatment of burns as silver owes the efficiency to restrain the colony formation and growth of multidrug-resistant pathogens (e.g., methicillin-resistant *Staphylococcus aureus,* gentamicin-resistant *Pseudomonas aeruginosa*) (Jain et al. 2009). Moreover, metallic NPs are effective against both gram-positive and gram-negative bacteria (Ge et al. 2014).

9.5 Conclusions and Future Perspectives

From the past few years, the field of molecular imaging has seen pace and advancements that are very crucial and impactful in terms of biomedical applications. We have seen the applicative properties and advantages of the NMs over a plethora of other previously used contrasting agents. These advanced types of contrasting agents have tried to fill the gap between the clinicians and disease diagnosis, imaging, and treatment. Novel NMs with unique properties have made them perfect candidates for unique and unparalleled imaging sensitivity, cell/tissue specificity, and increased tissue penetration and also making them multimodal contrasting agents. Since, with these pros and properties, the expectations for better molecular imaging-based NMs have increased exponentially. Besides these applications, the prime concerns about the biocompatibility and toxicity have been in the mainstream in using these NMs for molecular imaging. A proper balance between the NM safety, synthesis/upscaling, and also its clinical commercialization will make the NMs a superior for the bioimaging and delivery with high sensitivity and selectivity. Making these things possible in the near future will help in curtailing the flaws and also in achieving the towering expectations in NM-based molecular imaging.

There is a hasty increase in the newer and advanced NMs for molecular imaging, but the path for clinical trials remains relatively full of hurdles and an uphill task. For the proper clinical translation of these NMs in molecular imaging and diagnostics, a number of crucial challenges need to be addressed, viz., biocompatibility of NMs, toxicity concerns, targeting efficacy, and also their long-term stability in the physiological and biological milieu. Moreover, for the desired application, choosing a suitable NM as well as the imaging modality is of prime importance. Further, by the integration of theranostics and multimodal molecular imaging of different NMs, it is expected to have fierce developments and advancements in the NM-based bioimaging and is expected to be a driving force for the next generation of theranostic systems. With swift advancements, the bio- and cellular compatibility of the NMs for bioimaging has been increasing with great speed. The bioimaging NMs are becoming more and more enthusiastic because of the increased sophistication for multimodal molecular imaging and also theranostic applications. Once these molecular imaging-abled NMs become the model candidates for bioimaging, they create a space for further advancements for making them as the perfect molecules for the clinicians as well as the patients and more likely will become the indispensable medical tools.

For the researchers and the clinicians, it is very important to properly identify the NM of choice for conjugating it with a superior imaging modality for the optimization of molecular imaging-based diagnosis of diseases. With the wide tunability of the characteristic properties of the NMs, they have become even more superior in contrast, sensitivity, specificity, spatial resolution, etc. Further, with increased advancements on the NM-based diagnostics, there is a need of not only the comprehensive knowledge but also a surfeit of in vitro, in vivo, and clinical studies. The proper implication of all the knowledge and studies will ultimately make it possible to precisely and accurately image the reachable disease tissue sites. Despite these stunning and fascinating advances, much is to be done for achieving the better of applications and potentials of these NMs for imaging, diagnostics, and treatment of diseases.

Toxicity studies of the synthesized NMs should be carried out with ease prior to their use in in vivo clinical trials. The diagnosis should also accommodate the monitoring of therapeutic effect. Current research should be mainly focused toward providing a simple, noninvasive, economical, sensitive, specific, and quick diagnosis, making personalized treatment with patient compliance (Luk and Zhang 2014). Future advancements on IVDs should be focusing on reducing the complexities for patients by providing a system that can be more reliable, user-friendly, and portable. In current trends for NMs, VOC sensing has given the much-anticipated platform for noninvasive diagnosis where the biomarkers of various diseases are detected from the breath of the patient. Despite this noninvasive approach, these systems lack sensitivity and credibility (Kim et al. 2017a, b). The latest technological advancements in nanotechnology are being applied to such NMs for required sensitivity and specificity. Then, NM-based targeted delivery of therapeutic agents must be worthy of increased attention in the upcoming time that involves recognition of different cellular and molecular receptors in the targeted area.

Acknowledgment The authors would like to thank the Director of CSIR-IHBT for his constant support and encouragement. AA acknowledges the financial support from CSIR (MLP-201) and DST (GAP-0214; EMR/2016/003027). Pooja and SS would like to acknowledge CSIR for their financial assistance from MLP-0145 and MLP-0141, respectively. The CSIR-IHBT communication number of this manuscript is 4585.

References

Anirudhan TS, Alexander S, Lilly A (2014) Surface modified multiwalled carbon nanotube based molecularly imprinted polymer for the sensing of dopamine in real samples using potentiometric method. Polymer 55(19):4820–4831

Ashley J, Shukor Y, D'Aurelio R, Trinh L, Rodgers TL, Temblay J, Pleasants M, Tothill IE (2018) Synthesis of molecularly imprinted polymer nanoparticles for α-casein detection using surface Plasmon resonance as a Milk allergen sensor. ACS Sens 3(2):418–424

Ashton JR, West JL, Badea CT (2015) In vivo small animal micro-CT using nanoparticle contrast agents. Front Pharmacol 6:256

Badea CT, Athreya KK, Espinosa G (2012) Computed tomography imaging of primary lung cancer in mice using a liposomal-iodinated contrast agent. PLoS One 7(4):e34496

Bashir MR, Bhatti L, Marin D, Nelson RC (2015) Emerging applications for ferumoxytol as a contrast agent in MRI. J Magn Reson Imaging 41(4):884–898

Benito AB, Aiertza MK, Marradi M, Gil-Iceta L, Shekhter Zahavi T, Szczupak B, Jiménez-González M, Reese T, Scanziani E, Passoni L, Matteoli M (2016) Functional single-chain polymer nanoparticles: targeting and imaging pancreatic tumors in vivo. Biomacromolecules 17 (10):3213–3221

Bernd H, De Kerviler E, Gaillard S, Bonnemain B (2009) Safety and tolerability of ultrasmall superparamagnetic iron oxide contrast agent: comprehensive analysis of a clinical development program. Investig Radiol 44(6):336–342

Cai S, Liu X, Han Q, Qi C, Yang R, Wang C (2018) A novel strategy to construct supported Pd nanocomposites with synergistically enhanced catalytic performances. Nano Res 11 (6):3272–3281

Chen S, Svedendahl M, Van Duyne RP, Käll M (2011) Plasmon-enhanced colorimetric ELISA with single molecule sensitivity. Nano Lett 11(4):1826–1830

Chen G, Qiu H, Prasad PN, Chen X (2014) Upconversion nanoparticles: design, Nanochemistry, and applications in Theranostics. Chem Rev 114:5161–5214

Cheng SH, Lo LW (2011) Inorganic nanoparticles for enhanced photodynamic cancer therapy. Curr Drug Discov Technol 8(3):269–276

Chou SS, Kaehr B, Kim J, Foley BM, De M, Hopkins PE, Huang J, Brinker CJ, Dravid VP (2013) Chemically exfoliated MoS2 as near-infrared Photothermal agents. Angew Chem Int Ed 52 (15):4160–4164

Clawson C, Ton L, Aryal S, Fu V, Esener S, Zhang L (2011) Synthesis and characterization of lipid-polymer hybrid nanoparticles with pH-Triggered Poly(Ethylene Glycol) shedding. Langmuir 27:10556–10561

Colombeau L, Acherar S, Baros F (2016) Inorganic nanoparticles for photodynamic therapy. Top Curr Chem 370:113–134

Cormode DP, Naha PC, Fayad ZA (2014) Nanoparticle contrast agents for computed tomography: a focus on micelles. Contrast Media Mol Imaging 9(1):37–52

Curry T, Kopelman R, Shilo M, Popovtzer R (2014) Multifunctional theranostic gold nanoparticles for targeted CT imaging and photothermal therapy. Contrast Media Mol Imaging 9(1):53–61

Daldrup-Link HE (2017) Ten things you might not know about iron oxide nanoparticles. Radiology 284:616–629

De La Rica R, Stevens MM (2012) Plasmonic ELISA for the ultrasensitive detection of disease biomarkers with the naked eye. Nat Nanotechnol 7(12):821

Dolmans D, Fukumura D, Jain RK (2003) Photodynamic therapy for cancer. Nat Rev Cancer 3 (5):380–387

Dutz S, Hergt R (2014) Magnetic particle hyperthermia-a promising tumour therapy? Nanotechnology 25(45):452001

Ellis JE, Star A (2016) Carbon nanotube based gas sensors toward breath analysis. ChemPlusChem 81(12):1248–1265

Feldman EJ, Lancet JE, Kolitz JE, Ritchie EK, Roboz GJ, List AF, Allen SL, Asatiani E, Mayer LD, Swenson C (2011) First-in-man study of CPX-351: a liposomal carrier containing Cytarabine and Daunorubicin in a fixed 5:1 molar ratio for the treatment of relapsed and refractory acute myeloid Leukemia. J Clin Oncol 29(8):979–985

Feng T, Wang Y, Qiao X (2017) Recent advances of carbon nanotubes-based electrochemical Immunosensors for the detection of protein cancer biomarkers. Electroanalysis 29(3):662–675

Fu C, Yang W, Chen X, Evans DG (2009) Direct electrochemistry of glucose oxidase on a graphite nanosheet–Nafion composite film modified electrode. Electrochem Commun 11(5):997–1000

Gao Y, Hernandez C, Yuan HX (2017) Ultrasound molecular imaging of ovarian cancer with CA-125 targeted nanobubble contrast agents. Nanomedicine 13(7):2159–2168

Ge L, Li Q, Wang M, Ouyang J, Li X, Xing MMQ (2014) Nanosilver particles in medical applications: synthesis, performance, and toxicity. Int J Nanomedicine 9:2399–2407

Ghosal K, Sarkar K (2018) Biomedical applications of graphene nanomaterials and beyond. ACS Biomater Sci Eng 4(8):2653–2703

González-Gaitán C, Ruiz-Rosas R, Morallon E, Cazorla-Amorós D (2017) Effects of the surface chemistry and structure of carbon nanotubes on the coating of glucose oxidase and electrochemical biosensors performance. RSC Adv 7(43):26867–26878

Heesakkers RA, Jager GJ, Hövels AM (2009) Prostate cancer: detection of lymph node metastases outside the routine surgical area with ferumoxtran-10-enhanced MR imaging. Radiology 251 (2):408–414

Hirsch LR, Stafford RJ, Bankson JA, Sershen SR, Rivera B, Price RE, Hazle JD, Halas NJ, West JL (2003) Nanoshell- mediated near-infrared thermal therapy of tumors under magnetic resonance guidance. Proc Natl Acad Sci U S A 100(23):13549–13554

Hong EJ, Choi DG, Shim MS (2016) Targeted and effective photodynamic therapy for cancer using functionalized nanomaterials. Acta Pharm Sin B 6(4):297–307

Howes PD, Rana S, Stevens MM (2014) Plasmonic nanomaterials for biodiagnostics. Chem Soc Rev 43(11):3835–3853

Hu Z, Guan W, Wang W, Huang L, Tang X, Xu H, Xing H (2008) Synthesis of amphiphilic amino acid C60 derivatives and their protective effect on hydrogen peroxide-induced apoptosis in rat pheochromocytoma cells. Carbon 46(1):99–109

Hu Z, Guan W, Wang W, Zhu Z, Wang Y (2010a) Folacin C60 derivative exerts a protective activity against oxidative stress-induced apoptosis in rat pheochromocytoma cells. Bioorg Med Chem letters 20(14):4159–4162

Hu CMJ, Kaushal S, Cao HST, Aryal S, Sartor M, Esener S, Bouvet M, Zhang L (2010b) Half-antibody functionalized lipid-polymer hybrid nanoparticles for targeted drug delivery to carcinoembryonic antigen presenting pancreatic cancer cells. Mol Pharm 7:914–920

Hu CM, Fang RH, Luk BT, Zhang L (2014) Polymeric Nanotherapeutics: clinical development and advances in stealth functionalization strategies. Nanoscale 6:65–75

Huang XH, Neretina S, El-Sayed MA (2009) Gold Nanorods: from synthesis and properties to biological and biomedical applications. Adv Mater 21(48):4880–4910

Inci F, Tokel O, Wang S, Gurkan UA, Tasoglu S, Kuritzkes DR, Demirci U (2013) Nanoplasmonic quantitative detection of intact viruses from unprocessed whole blood. ACS Nano 7 (6):4733–4745

Jain J, Arora S, Rajwade JM, Omray P, Khandelwal S, Paknikar KM (2009) Silver nanoparticles in therapeutics: development of an antimicrobial gel formulation for topical use. Mol Pharm 6 (5):1388–1401

Kamaly N, Xiao ZY, Valencia PM, Radovic-Moreno AF, Farokhzad OC (2012) Targeted polymeric therapeutic nanoparticles: design, development and clinical translation. Chem Soc Rev 41 (7):2971–3010

Kang T, Yoo SM, Yoon I, Lee SY, Kim B (2010) Patterned multiplex pathogen DNA detection by Au particle-on-wire SERS sensor. Nano Lett 10(4):1189–1193

Khan R, Pal M, Kuzikov AV, Bulko T, Suprun EV, Shumyantseva VV (2016) Impedimetric immunosensor for detection of cardiovascular disorder risk biomarker. Mater Sci Eng C 68:52–58
Kim K, Lee M, Park H, Kim JH, Kim S, Chung H, Kwon IC (2006) Cell-permeable and biocompatible polymeric nanoparticles for apoptosis imaging. J Am Chem Soc 128 (11):3490–3491
Kim MS, Kweon SH, Cho S, An SSA, Kim MI, Doh J, Lee J (2017a) Pt-decorated magnetic nanozymes for facile and sensitive point-of-care bioassay. ACS Appl Mater Interfaces 9 (40):35133–35140
Kim SJ, Choi SJ, Jang JS, Cho HJ, Kim ID (2017b) Innovative nanosensor for disease diagnosis. Acc Chem Res 50(7):1587–1596
Kumar A, Kumar S, Rhim WK, Kim GH, Nam JM (2014) Oxidative nanopeeling chemistry-based synthesis and photodynamic and photothermal therapeutic applications of plasmonic core-petal nanostructures. J Am Chem Soc 136(46):16317–16325
Lalwani G, Cai X, Nie L, Wang LV, Sitharaman B (2013) Graphene-based contrast agents for photoacoustic and thermoacoustic tomography. Photo-Dermatology 1(3–4):62–67
Li J, Zhen X, Lyu Y, Jiang Y, Huang J, Pu K (2018) Cell membrane coated semiconducting polymer nanoparticles for enhanced multimodal cancer phototheranostics. ACS Nano 12 (8):8520–8530
Lim DK, Jeon KS, Kim HM, Nam JM, Suh YD (2010) Nanogap-engineerable Raman-active nanodumbbells for single-molecule detection. Nat Mater 9(1):60
Liu CH, Ren JQ, Yang J (2009) DNA-based MRI probes for specific detection of chronic exposure to amphetamine in living brains. J Neurosci 29(34):10663–10670
Liu Q, Wei L, Wang J, Peng F, Luo D, Cui R, Yang J (2012) Cell imaging by graphene oxide based on surface enhanced Raman scattering. Nanoscale 4(22):7084–7089
Luk BT, Zhang L (2014) Current advances in polymer-based nanotheranostics for cancer treatment and diagnosis. ACS Appl Mater Interfaces 6(24):21859–21873
Lv R, Wang D, Xiao L (2017) Stable ICG-loaded upconversion nanoparticles: silica core/shell theranostic nanoplatform for dual-modal upconversion and photoacoustic imaging together with photothermal therapy. Sci Rep 7:15753
Ma L, Tu C, Le P, Chitoor S, Lim SJ, Zahid MU, Smith AM (2016) Multidentate polymer coatings for compact and homogeneous quantum dots with efficient bioconjugation. J Am Chem Soc 138 (10):3382–3394
Maji SK, Sreejith S, Joseph J, Lin M, He T, Tong Y, Sun H, Yu SW, Zhao Y (2014) Upconversion nanoparticles as a contrast agent for photoacoustic imaging in live mice. Adv Mater 26 (32):5633–5638
Manna B, Raj CR (2016) Covalent functionalization and electrochemical tuning of reduced graphene oxide for the bioelectrocatalytic sensing of serum lactate. J Mater Chem B 4 (26):4585–4593
Markwalter CF, Kantor AG, Moore CP, Richardson KA, Wright DW (2018) Inorganic complexes and metal-based nanomaterials for infectious disease diagnostics. Chem Rev 119(2):1456–1518
McCarthy JR, Weissleder R (2008) Multifunctional magnetic nanoparticles for targeted imaging and therapy. Adv Drug Deliv Rev 60(11):1241–1251
Mitragotri S, Anderson DG, Chen X, Chow EK, Ho D, Kabanov AV, Shi J (2015) Accelerating the translation of nanomaterials in biomedicine. ACS Nano 9(7):6644–6654
Mody VV, Nounou MI, Bikram M (2009) Novel nanomedicine-based MRI contrast agents for gynecological malignancies. Adv Drug Deliv Rev 61(10):795–807
Moen MD, Lyseng-Williamson KA, Scott LJ (2009) Liposomal Amphotericin B a review of its use as empirical therapy in febrile neutropenia and in the treatment of invasive fungal infections. Drugs 69(3):361–392
Morgan J, Oseroff AR (2001) Mitochondria-based photodynamic anti-cancer therapy. Adv Drug Deliv Rev 49(1–2):71–86

Navamani J, Palanisamy R, Gurusamy R, Ramasamy M, Arumugam S (2012) Development of nanoprobe for the determination of blood cholesterol. J Biosens Bioelectron 3:1–8

Oh WK, Jeong YS, Kim S, Jang J (2012) Fluorescent polymer nanoparticle for selective sensing of intracellular hydrogen peroxide. ACS Nano 6(10):8516–8524

Park JY, Chang Y, Lee GH (2015) Multi-modal imaging and cancer therapy using lanthanide oxide nanoparticles: current status and perspectives. Curr Med Chem 22(5):569–581

Patil AV, Fernandes FB, Bueno PR, Davis JJ (2015) Graphene-based protein biomarker detection. Bioanalysis 7(6):725–742

Petros RA, DeSimone JM (2010) Strategies in the design of nanoparticles for therapeutic applications. Nat Rev Drug Discovery 9:615–627

Ren J, Chen YI, Liu CH (2016) Noninvasive tracking of gene transcript and neuroprotection after gene therapy. Gene Ther 23(1):1–9

Rief M, Wagner M, Franiel T (2009) Detection of focal liver lesions in unenhanced and ferucarbotran-enhanced magnetic resonance imaging: a comparison of T2-weighted breath-hold and respiratory-triggered sequences. Magn Reson Imaging 27(9):1223–1229

Robinson JT, Tabakman SM, Liang YY, Wang HL, Sanchez Casalongue H, Vinh D, Dai HJ (2011) Ultrasmall reduced graphene oxide with high near-infrared absorbance for photothermal therapy. J Am Chem Soc 133(17):6825–6831

Rodríguez-Lorenzo L, de La Rica R, Álvarez-Puebla RA, Liz-Marzán LM, Stevens MM (2012) Plasmonic nanosensors with inverse sensitivity by means of enzyme-guided crystal growth. Nat Mater 11(7):604

Rzigalinski BA, Strobl JS (2009) Cadmium-containing nanoparticles: perspectives on pharmacology and toxicology of quantum dots. Toxicol Appl Pharmacol 238(3):280–288

Segev-Bar M, Haick H (2013) Flexible sensors based on nanoparticles. ACS Nano 7 (10):8366–8378

Shan C, Yang H, Han D, Zhang Q, Ivaska A, Niu L (2010) Graphene/AuNPs/chitosan nanocomposites film for glucose biosensing. Biosens Bioelectron 25(5):1070–1074

Shashkov EV, Everts M, Galanzha EI, Zharov VP (2008) Quantum dots as multimodal photoacoustic and photothermal contrast agents. Nano Lett 8:3953–3958

Shu C, Corwin FD, Zhang J, Chen Z, Reid JE, Sun M, Esker AR (2009) Facile preparation of a new gadofullerene-based magnetic resonance imaging contrast agent with high 1H relaxivity. Bioconjug Chem 20(6):1186–1193

Tang Z, Liu Y, Duan Y (2015) Development of solid-phase microextraction fibers based on multi-walled carbon nanotubes for pre-concentration and analysis of alkanes in human breath. J Chromatogr A 1425:34–41

Thomas R, Park IK, Jeong YY (2013) Magnetic iron oxide nanoparticles for multimodal imaging and therapy of cancer. Int J Mol Sci 14(8):15910–15930

Tian J, Huang J, Zhao Y, Zhao S (2012) Electrochemical immunosensor for prostate-specific antigen using a glassy carbon electrode modified with a nanocomposite containing gold nanoparticles supported with starch-functionalized multi-walled carbon nanotubes. Microchim Acta 178(1–2):81–88

Ulissi ZW, Sen F, Gong X, Sen S, Iverson N, Boghossian AA, Strano MS (2014) Spatiotemporal intracellular nitric oxide signaling captured using internalized, near-infrared fluorescent carbon nanotube nanosensors. Nano Lett 14(8):4887–4894

Verma S, Watt GM, Mai Z, Hasan T (2007) Strategies for enhanced photodynamic therapy effects. Photochem Photobiol 83(5):996–1005

Vilela P, El-Sagheer A, Millar TM, Brown T, Muskens OL, Kanaras AG (2016) Graphene oxide-upconversion nanoparticle based optical sensors for targeted detection of mRNA biomarkers present in Alzheimer's disease and prostate cancer. ACS sensors 2(1):52–56

Wang AZ, Langer R, Farokhzad OC (2012) Nanoparticle delivery of cancer drugs. Annu Rev Med 63:185–198

Wang Z, He X, Yong T, Miao Y, Zhang C, Tang BZ (2020) Multicolor Tunable polymeric nanoparticle from Tetraphenylethylene-cage for temperature sensing in living cells. J Am Chem Soc 115(21):11718–11940

Weber J, Beard PC, Bohndiek SE (2016) Contrast agents for molecular photoacoustic imaging. Nat Methods 13:639–650

Wolfbeis OS (2015) An overview of nanoparticles commonly used in fluorescent bioimaging. Chem Soc Rev 44:4743–4768

Wu X, Xu L, Liu L, Ma W, Yin H, Kuang H, Kotov NA (2013) Unexpected chirality of nanoparticle dimers and ultrasensitive chiroplasmonic bioanalysis. J Am Chem Soc 135 (49):18629–18636

Xi Z, Ye H, Xia X (2018) Engineered noble-metal nanostructures for in vitro diagnostics. Chem Mater 30(23):8391–8414

Xu L, Yan W, Ma W, Kuang H, Wu X, Liu L, Xu C (2015) SERS encoded silver pyramids for attomolar detection of multiplexed disease biomarkers. Adv Mater 27(10):1706–1711

Xu G, Lin G, Lin S (2016) The reproductive toxicity of CdSe/ZnS quantum dots on the in vivo ovarian function and in vitro fertilization. Sci Rep 6:37677

Xu S, Ouyang W, Xie P, Lin Y, Qiu B, Lin Z, Guo L (2017a) Highly uniform gold nanobipyramids for ultrasensitive colorimetric detection of influenza virus. Anal Chem 89(3):1617–1623

Xu H, Wang L, Fan C (2017b) Bioanalysis and bioimaging with fluorescent conjugated polymers and conjugated polymer nanoparticles. In: Functional nanoparticles for bioanalysis, Nanomedicine, and bioelectronic devices volume 1. American Chemical Society, Washington, DC, pp 81–117

Yang X, Yu Y, Gao Z (2014) A highly sensitive plasmonic DNA assay based on triangular silver nanoprism etching. ACS Nano 8(5):4902–4907

Zackrisson S, van de Ven S, Gambhir SS (2014) Light in and sound out: emerging translational strategies for photoacoustic imaging. Cancer Res 74:979–1004

Zhan N, Palui G, Safi M, Ji X, Mattoussi H (2013) Multidentate zwitterionic ligands provide compact and highly biocompatible quantum dots. J Am Chem Soc 135(37):13786–13795

Zhou W, Gao X, Liu D, Chen X (2015) Gold nanoparticles for in vitro diagnostics. Chem Rev 115 (19):10575–10636

The manufacturer's authorised representative in the EU is Springer
Nature Customer Service Centre GmbH, Europaplatz 3, 69115 Heidelberg,
Germany. If you have any concerns regarding our products, please
contact ProductSafety@springernature.com

Printed and bound by CPI Group (UK) Ltd, Croydon, CR0 4YY
30/09/2026
02225317-0001